MW01674174

Organizing and Training Airfield Operations Capabilities for Emerging Expeditionary Operations

Potential Courses of Action

JOHN A. AUSINK, DWAYNE M. BUTLER, KRISTIN F. LYNCH, JOHN G. DREW, SARAH W. DENTON, MONICA RICO, IGNACIO A. LARA, WILL SHUMATE, ALBERT M. ESPOSITO

Prepared for the Department of the Air Force
Approved for public release; distribution unlimited

RAND PROJECT AIR FORCE

For more information on this publication, visit **www.rand.org/t/RRA990-2**.

Published by the RAND Corporation, Santa Monica, Calif.

Library of Congress Cataloging-in-Publication Data is available for this publication.

ISBN: 978-1-9774-0967-6

Cover: Airman 1st Class Kevin Tanenbaum/U.S. Air Force.

About This Report

The research reported here was commissioned by the Office of the Deputy Chief of Staff for Operations, Headquarters U.S. Air Force (USAF), and conducted within the Workforce, Development and Health Program of RAND Project Air Force as part of a fiscal year 2021 project, "Organizing and Training Airfield Operations Capabilities for Expeditionary Operations." The objective of the project was to help enhance the ability of the USAF Airfield Operations career fields to respond to the demands created by the Department of Defense's Dynamic Force Employment concept and the USAF's implementation of the Agile Combat Employment concept.

RAND Project AIR FORCE

RAND Project AIR FORCE (PAF), a division of the RAND Corporation, is the Department of the Air Force's (DAF's) federally funded research and development center for studies and analyses, supporting both the United States Air Force and the United States Space Force. PAF provides the DAF with independent analyses of policy alternatives affecting the development, employment, combat readiness, and support of current and future air, space, and cyber forces. Research is conducted in four programs: Strategy and Doctrine; Force Modernization and Employment; Resource Management; and Workforce, Development, and Health. The research reported here was prepared under contract FA7014-16-D-1000.

Additional information about PAF is available on our website:

www.rand.org/paf/

This report documents work originally shared with the DAF on September 30, 2021. The draft report, issued on September 30, 2021, was reviewed by formal peer reviewers and DAF subject-matter experts.

Acknowledgments

The authors of this report are grateful to the project sponsor, Lt Gen Joseph T. Guastella Jr., Air Force Deputy Chief of Staff for Operations (AF/A3) for his sponsorship of this research.[1] Brig. Gen. David Mineau, Director of Current Operations (AF/A3O) and Col Daniel C. Furleigh (AF/A3OJ) were our initial contacts for the project, and we appreciate their support. We are especially grateful to our action officer, Lt Col Billy R. Wilson, Branch Chief for Airfield Operations Force Development, for his tremendous assistance in providing interview contacts

[1] All ranks and offices current as of publication.

and leads for data and documents. In addition, he invited our research team to participate in the October 2020 Airfield Operations Executive Council meeting and in the ongoing Airfield Operations Vision and Bridging strategy working groups, which gave us important insights into the challenges faced by the Airfield Operations community. Numerous individuals from Headquarters Air Force, combatant commands, major commands, the Air National Guard, the Air Force Reserve, and other organizations were very generous with their time by granting interviews and providing data and documents. The project would not have been possible without their assistance.

We thank RAND colleague Pat Mills, who served as a project advisor, and Barbara Bicksler, who was our communications analyst. Their feedback and advice on improving the presentation of the report was invaluable.

Summary

Issue

Airfield operations (AO) officer and enlisted career fields provide air traffic control; airfield management; and radar, airfield, and weather systems services for the U.S. Air Force (USAF), in-garrison and deployed, during peacetime and contingencies. The AO community faces three challenges in adjusting to a new strategic environment. First is the introduction of the Dynamic Force Employment (DFE) concept involving a transition from centralized, unhardened infrastructure to smaller, dispersed, resilient, adaptive basing, which could increase the demands for AO services. Second, the USAF is developing force employment ideas that are grouped under the term Agile Combat Employment (ACE), which will involve quickly dispersing and clustering forces at different locations in a way that will allow the United States to increase the survivability of its forces.[2] Finally, to reduce the strain and prepare to respond to near-peer adversaries while incorporating DFE and ACE concepts, the USAF is implementing a new Air Force Force Generation approach (AFFORGEN).

As these concepts mature and evolve, it is important for the USAF to determine whether the demands for AO in a changing strategic environment can be satisfied. The objective of this project is to identify ways to enhance the ability of the AO career fields to respond to the needs created by the DFE concept and the USAF implementation of ACE.

Approach

The research team on this project used three research streams. A review of USAF and joint documents related to the AO career fields provided background for AO doctrine and missions. Interviews, informal discussions, and other interactions with a variety of subject-matter experts illuminated current challenges for accomplishing AO missions and highlighted the potential impact of new demands on AO that could result from DFE and ACE approaches. Finally, using AO personnel data and current approaches to ACE concepts, simple R-based tools were developed to analyze the ability of AO personnel to meet future demands.

Key Findings

Key findings fall in two areas:

[2] The FY 22 Department of the Air Force Posture Statement states "This new method of operating will allow the United States to confuse the enemy and strike at a time and place of our choosing with minimal risk. The ability of ACE to sustain combat operations through persistent logistics has already been validated through multiple exercises across both the Pacific and European theaters" (Roth, Brown, and Raymond, 2021, p. 13).

- The supply and demand for AO capabilities:
  - Viewing AO capabilities through supply and demand lenses is necessary for the AO community to organize, train, and equip to satisfy expected mission demand.
  - Even if all AO demands were well understood, there could be challenges in satisfying them. The USAF has four "special" sources of airfield operations expertise—Operational Support Squadrons, contingency response groups, the 53rd Air Traffic Control Squadron, and ten Air National Guard air traffic control squadrons. The relationships among these organizations are not well understood, nor is it clear how their services are requested and provided.
  - AO equipment is outdated and may not be able to meet the requirements for hub-and-spoke operations as envisioned by ACE concepts of operations.
- Management of AO resources:
  - New approaches to managing personnel resources are needed.
  - The new AFFORGEN model may provide an opportunity to posture AO capabilities from all AO sources to satisfy emerging DFE and ACE requirements.

Recommendations

Table S.1 shows recommended courses of action associated with our key findings.

Table S.1. Recommended Courses of Action

Observation	Recommended Course of Action
Supply of and Demand for AO Capabilities	
Viewing AO capabilities through supply and demand lenses is important.	*Designate a champion to monitor overall demand* The champion or proponent—an office or organization—would track, quantify, analyze, prioritize, and publish demands for the AO community.
Even if all AO demands were well understood, it is not always clear how decisions are made to align available supply with demand.	*Develop analytic tools and capability* Until the USAF understands the overall demands—as monitored by a designated champion—it will struggle to align available supply to the demands. *Identify decision authority* A decisionmaking authority needs to be established to determine the priority for filling AO capability needs. We recommend that the Headquarters Air Force Functional Area Manager fill this role.
Deployable AO equipment is outdated and may not be appropriate for hub-and-spoke operations.	*Align active-duty units with Air National Guard air traffic control units to make more efficient use of existing equipment, and coordinate with the Air Force Flight Standards Agency to procure common equipment across services* Aligning equipment across components would improve the utilization of current equipment; new deployable equipment will still be needed to provide required capabilities of the hub-and-spoke concept within ACE.
Management of AO Resources	
New approaches to managing personnel resources are needed.	*Consider establishing active-associate units at high-priority locations in the continental United States (CONUS)* The Air Force should explore approaches to using Air National Guard personnel to replace active-duty personnel when they are deployed from key operating locations, such as Air Mobility Command bases. Innovative approaches to using civilian personnel and contractors at CONUS locations during contingencies should also be considered.
The new AFFORGEN model may provide an opportunity to leverage AO capabilities to satisfy expeditionary requirements.	*Fit the AO community into the AFFORGEN model as a capability* Within AFFORGEN, supply prioritization and equipping strategies (in previous recommendations) can be better leveraged to posture the AO community for contingency operations that incorporate DFE and ACE.

Contents

Figures and Tables

Figures

Tables

1. Introduction

Airfield operations (AO) in the U.S. Air Force (USAF) include air traffic control (ATC); airfield management; and radar, airfield, and weather systems (RAWS) maintenance. It is USAF policy to sustain resources that provide a "safe, operationally relevant flying environment that promotes combat readiness in compliance with applicable laws and regulations" (Air Force Policy Directive [AFPD] 13-2, 2019, p. 2). To accomplish this, the USAF must

- provide airspace, aviation support facilities, equipment, and services needed to accomplish the flying mission and provide complementary services that are integrated into the national airspace system
- "Staff, operate, and maintain airfield operations assets in peacetime for the purpose of directly supporting the expeditionary nature of USAF contingency operations in terms of readiness and deployability" (AFPD 13-2, 2019, p. 2).

AO personnel in four career fields[3] support flying operations during peacetime at home stations in the continental United States (CONUS) and at locations overseas. Contingency operations—"situations requiring military operations in response to natural disasters, terrorists, subversives, or as otherwise directed by appropriate authority to protect United States interests" (Air Force Manual [AFMAN] 13-204, Vol. 1, 2020a, p. 6)— may take place at home stations in CONUS or outside the continental United States (OCONUS) but also at locations that are newly established in response to contingency requirements, such as forward operating bases (FOBs) and forward operating sites (FOSs).

A Changing Strategic Environment

The 2017 *National Security Strategy* stated that the Department of Defense (DoD) "must develop new operational concepts and capabilities to win without assured dominance in air, maritime, land, space, and cyberspace domains, including against those operating below the level of conventional military conflict" (Trump, 2017, p. 29). One new operational concept in response to this direction is Dynamic Force Employment (DFE), a still-evolving approach to force posture and employment that is meant to provide "options for proactive and scalable employment of the Joint Force" (Mattis, 2018, p. 7). A component of this approach will be a transition from centralized, unhardened infrastructure to smaller, dispersed, resilient, adaptive basing (Mattis, 2018, p. 6).

In the meantime, the USAF is developing force employment ideas that are grouped under the term Agile Combat Employment (ACE). According to the Chief of Staff of the Air Force,

---

[3] The officer career field for AO is 13M, airfield operations officer. Three enlisted career fields are 1C1XX (air traffic control), 1C7XX (airfield management), and 1C8XX (RAWS services).

> ACE is the ability to quickly disperse and cluster forces to a cooperative security location and conduct operations across all domains with minimal disruption, while maintaining operational flexibility. This new method of operating will allow the United States to confuse the enemy and strike at a time and place of our choosing with minimal risk (Roth, Brown, and Raymond, 2021).

The demands for USAF missions for the past two decades have strained its forces and had a negative effect on readiness (Roth, Brown, and Raymond, 2021, p. 12). To reduce the strain and prepare to respond to near-peer adversaries while incorporating DFE and ACE concepts, the USAF is implementing a new Air Force Force Generation (AFFORGEN) model that is

> focused on building and sustaining long-term, high-end readiness. Our goal is to more effectively articulate readiness impacts and capacity limits, and instill discipline into the system (Roth, Brown, and Raymond, 2021, p. 12).

The transition to more scalable forces under DFE and the approach to agility and posture under ACE will have an impact on the demand for AO personnel, and the AFFORGEN concept could affect how their services are supplied. As these concepts mature and evolve, it is important for the USAF to determine whether the demands for AO in a changing strategic environment can be satisfied.

Research Objective and Approach

Lt Gen Joseph T. Guastella Jr., Deputy Chief of Staff for Operations, Headquarters USAF (AF/A3), desires to meet the *National Security Strategy* mandate to restore the Air Force's ability to produce innovative capabilities, maintain readiness, and "grow the size of the force so that it is capable of operating at sufficient scale and for ample duration to win across a range of scenarios" (Trump, 2017, p. 28). With a focus on the AO component of this mandate, the objective of this project is to help enhance the ability of the AO community to respond to the needs created by the DFE concept and the USAF implementation of ACE.

To accomplish this objective, the study team examined

- the current AO "ecosystem," including personnel and missions
- the implications of DFE, ACE, and AFFORGEN for the AO community
- the challenges that the AO community may face in accomplishing current and future missions.

The results of these efforts culminated in courses of action that the USAF could pursue to address the identified challenges.

We were fortunate that, shortly after the project was approved, the USAF conducted its annual Airfield Operations Executive Council meeting on October 26–29, 2020, and RAND Project AIR FORCE (PAF) was invited to participate.[4] During the executive council meeting, the

[4] Because of COVID-19 limitations, only one member of the research team was allowed to attend in person. Other research team members participated virtually.

study team heard presentations from subject-matter experts (SMEs) on the perceptions of existing challenges for the AO career fields and heard suggestions for topics that should be explored as part of our research. This helped us develop our research approach:

- a literature review of Federal Aviation Administration (FAA), DoD, and Air Force documents related to AO; joint and Air Force doctrine; government documents related to military strategy—particularly those related to the implementation of DFE; Air Force documents describing ACE; and recent briefings describing AFFORGEN concepts
- semistructured interviews with SMEs that addressed standard questions about operations at in-garrison and deployed locations and how these differed during peacetime and contingency operations. Questions were designed from a doctrine, organization, training, materiel, leadership and education, personnel, facilities, and policy (DOTmLPF-P) perspective to help develop approaches to addressing challenges faced by the AO community.
- an analysis of manpower and personnel data to examine assigned and authorized active, reserve, and civilian component personnel in the AO career fields and to quantify requirements for these personnel
- modeling through a proof-of-concept tool to explore the impact of changes in the demand for AO personnel that might result from the implementation of DFE and ACE concepts
- RAND team participation in ongoing "Vision and Bridging Strategy" group meetings during which AO problems and proposed solutions were discussed.[5]

Structure of This Report

The remainder of this report presents the results of our research. Chapter 2 describes responsibilities of AO personnel in more detail, displays data on the sizes of different career fields and their distribution among Air Force components, describes important equipment required for AO tasks, and details the importance of clarifying the unusual demands that are placed on AO personnel.

Chapter 3 presents more details on DFE, ACE, and AFFORGEN concepts and the impact that they may have on AO responsibilities.

In Chapter 4, we use the DOTmLPF-P framework to summarize AO challenges that were described by those we interviewed and by participants in the Vision and Bridging Strategy sessions. In Chapter 5, we build on the discussion of Chapter 3 and the awareness of the challenges highlighted in Chapter 4 to quantify potential personnel shortages arising from the implementation of ACE concepts and describe proof-of-concept tools that can assist in the management of AO personnel.

Chapter 6 concludes with overall observations and suggested courses of action to respond to the challenges of a new strategic environment with DFE and ACE.

---

[5] Working sessions included action officers and senior steering group levels from all the major commands. The RAND PAF project team participated in the nine working group sessions and executive steering group sessions and more than ten postsession engagements from November 2020 through May 2021.

Appendix A lists the organizations interviewed for this project and describes the qualitative approach used to analyze interview responses. Appendix B provides additional information for the supply and demand framework that was developed through our participation in the Vision and Bridging Strategy sessions. Appendix C describes our personnel data sources and the AO tool we developed to examine the supply and demand of AO personnel.

2. Airfield Operations Personnel and Equipment

One of the challenges faced by the AO community is that functions performed by AO personnel at their home stations must continue to be accomplished even when some of a unit's AO personnel are deployed. For example, ATC operations do not stop at a CONUS mobility aircraft wing when some of the unit's ATC personnel are deployed to manage air traffic at a contingency location. Thus, as changes in national security strategy increase the demand for the deployment of AO capabilities, the Air Force will have to balance the need to maintain operations at home stations while providing personnel for deployment requirements.

This chapter briefly describes functions performed by AO personnel; the supply of, and some of the current demand for, personnel in the AO career fields; and some of the equipment needs of personnel in the career field.

Responsibilities of Airfield Operations Personnel

As denoted in the previous chapter, the enlisted and officer AO career fields perform functions in three areas—ATC, airfield management, and RAWS; these are described in this section.[6]

Air Traffic Control

The FAA defines *air traffic control* as a "service operated by appropriate authority to promote the safe, orderly and expeditious flow of air traffic" (FAA, 2019, Pilot Controller Glossary A-5). Duties of Air Force ATC personnel include controlling live or simulated traffic, monitoring a trainee controlling live or simulated traffic, issuing ATC movement and control messages, monitoring a trainee issuing ATC movement and control messages, and conducting duties as a watch supervisor (AFM 13-204, Vol. 1, 2020a, p. 55). The Air Force enlisted career field associated with ATC is 1C1X1.[7]

---

[6] The Air Force and DoD occupy "a unique, three-part role associated with the National Airspace System": The Air Force is a user of national airspace; it provides services through ATC and maintenance of navigational aids; and it is a regulatory authority, with the authority to certify aircraft airmen and airfields (AFPD 13-2, 2019). The Air Force has "a vested interest in ensuring and enhancing [its] ability to operate seamlessly with civil aviation in both national and international airspace" (AFPD 13-2, 2019, p. 1).

[7] The first character is career group (1 = operations), second character is career field (C is "command and control systems operations"), third character combined with first and second gives the career field subgroup, and the fourth character is skill level (1—Helper, 3—Apprentice, 5—Journeyman, 7—Craftsman, 9—Superintendent, 0—Chief Enlisted Manager). The fifth character combined with the others specifies the Air Force specialty code (AFSC) (Headquarters Air Force Personnel Center, 2020, and Air Force Instruction [AFI] 36-2101, 2006).

Airfield Management

Airfield management involves dealing with activities related to the airfield itself, such as construction and repairs; inspections for safety and compliance with airfield planning and design criteria; activities related to air travel, such as providing information for flight information publications, aeronautical charts, and navigational aid status; and activities related to aircraft operations, such as refueling and transient aircraft maintenance (AFM 13-204, Vol. 1, 2020a). These activities are accomplished by Air Force enlisted personnel in the 1C7X1 career field.

Radar, Airfield, and Weather System Services

RAWS responsibilities include the deployment, installation, and maintenance of ATC, weather, radar, navigation, and ground-to-air radio systems. "RAWS may be required to maintain entire facilities, subsystems, or individual services or equipment that assist in the safe and expedient movement of air traffic throughout the National Airspace System (NAS) as well as providing support to the National Weather Service (NWS)" (U.S. Air Force, 2018, p. 13). RAWS responsibilities are carried out by enlisted personnel in the 1C8X3 career field, which is relatively new. In 2006, the meteorological and navigation systems career field and a portion of the ground radio career field were combined to create the Airfield Systems career field. Then, in 2017, Airfield Systems was combined with Ground Radar Maintenance to create RAWS (Bryan Wilson, 2020).

Airfield Operations Officer

Finally, the officer career field for airfield management, 13MX (airfield operations officer), includes responsibilities for leading AO functions of ATC, airfield management, air traffic control and landing systems (ATCALS), and airspace management within the National Airspace System, International Civil Aviation Organization, and combat environments (U.S. Air Force, 2016, p. 8).

The Supply of Airfield Operations Personnel

The Air Force has about 6,000 permanent party (permanently assigned to a location) military personnel in the active, guard, and reserve components whose primary AFSCs are in AO. When students, operational students, transients, and "holdees" are included,[8] the total is about 7,800. Table 2.1 shows how the *permanent party* personnel in the active and guard components are

[8] *Students* are personnel whose current duty is being a student (such as attending a technical school). *Operational students* are rated personnel going to their initial full-time unit or back for requalification training. Test Pilot School is included in this category. *Transients* are personnel who are in transit (relocating to a new station or location). *Holdees* are prisoners.

distributed among the four AO career fields.[9] Of note, the officer AO career field is quite small, with just over 300 *total* officers in the active Air Force. The Air Force also employs about 1,500 government civilians in the air traffic controller (GS 2152) and transportation operations (GS 2150) career fields.[10]

Table 2.1. Airfield Operations Career Field Populations

	13M	1C1	1C7	1C8	Military Total	Civilian 2150	Civilian 2152	Civilian Total
Active	272	2,208	780	1,231	4,491	545	678	1,223
Air National Guard	42	395	320	676	1,433		1	1
Air Force Reserve	16	13	19		48	79	133	212
Unknown						46	23	69
Total	330	2,616	1,119	1,907	5,974			1,505

SOURCE: Military Personnel Data System data from February 2021. See Appendix C for details on how the dataset was created. Air Force Reserve (AFR) officer numbers are from a Unit Manpower Document provided by the AFR. "Unknown" civilian positions had blank entries for component in our database.
NOTE: Active personnel are permanent party (excluding students, ops students, holdees, and transients).

Presentations during the October 2020 meeting of the Airfield Operations Executive Council showed that all AO career fields were well-populated. The career fields were close to or exceeded 100 percent of their *total* authorized number of personnel (though as discussed later in this chapter, there is concern about the currency of the number of authorized positions). However, the distribution across grades did not always align with authorizations. For example, air traffic controllers had only 66 percent of their authorized E5s, and airfield managers had only 74 percent of their authorized E5s (Billy Wilson, 2020).

In addition to the AO personnel listed in Table 2.1, the Air Force supplements some of its AO personnel with contractors. For example, over 350 AO billets are required in U.S. Air Forces Central every six months, and 175 to 200 of these billets are filled with contractors who are paid for with overseas contingency operations funding.[11]

Distribution of Airfield Operations Personnel Among Units

Since the Air Force is concerned about meeting future deployment demands, it is useful to look at the distribution of AO personnel among four specific unit types in which individuals

[9] See Appendix C for a description of the databases used in this report.

[10] Civilians are in positions labeled with primary AFSC designations. There are about 130 in AO (13M) jobs, 694 in ATC (1C1X) jobs, and 161 in airfield management (1C7) jobs.

[11] Interview with representatives from U.S. Air Forces Central Command, November 18, 2020.

regularly exercise the ATC, airfield management, and RAWS skills important for contingency locations and which help illuminate the personnel who might be leveraged if contingency requirements increase: Operations Support Squadrons (OSSs), Contingency Response Groups and squadrons (CRGs and CRSs), Air Traffic Control Squadrons (ATCSs), and Regional Maintenance Centers (RMCs).

Table 2.2 summarizes the active-duty and Air National Guard (ANG) permanent party personnel in these four unit types: 77 percent of active-duty personnel and 59 percent of ANG permanent party personnel in the air operations career fields are assigned to these four unit types. The responsibilities of these units are described in the following sections.

Table 2.2. Active and Guard Military Permanent Party Personnel in Four Unit Types

	Number of Squadrons	13M	1C1	1C7	1C8	Total
Active duty						
OSS	67	143	1,812	616	578	3,149
ATC	1	4	10		88	102
CRG/S*	13	11	25	17	13	66
RMC	3				38	38
Total		158	1,864	646	778	**3,355**
Air National Guard						
OSS	72	0	7	257	0	264
ATC	10	22	327	2	226	577
Total						**841**

*Four CRGs, three CRS Support Squadrons, six CRSs.

Operations Support Squadrons

For an Air Force base with an airfield, the Airfield Operations Flight of the OSS provides airfield management and air traffic operations services to the base flying wing's transient and civil users. Typical flights include a RAWS section for maintaining ATC and airfield equipment (AFI 38-101, 2011, p. 91). There are about 67 active-duty OSSs.

Contingency Response Groups and Contingency Response Squadrons

Contingency response forces are meant to be the USAF's first responders. "They must have the ability to rapidly deploy, assess and open a location, and conduct initial operations immediately upon arrival in a variety of weather and threat environments from permissive to contested" (AFI 10-202, 2020, p. 8). CRGs are home station units that are organized to open forward contingency air bases and conduct air support operations anywhere in the world. CRGs are meant to "ensure mission-ready airfield assessment teams, airfield operations, C2 [command and control], aerial port, quick-turn aircraft maintenance, weather, intelligence, ATC, security forces, finance, fuels, supply, legal and contracting personnel are available to project and sustain combat forces worldwide" (AFI 10-202, 2020, p. 93). The Air Force has four active-duty CRGs:

two CRGs are under Air Mobility Command (AMC), one is under U.S. Air Forces in Europe (USAFE), and one is under Pacific Air Forces (PACAF). There is one CRG in the ANG.[12]

Air Traffic Control Squadrons

The 53rd ATCS is the only active-duty USAF ATCS. Its mission is to provide deployable ATCALS to combatant commanders when they need to establish a new airfield, project power into contested areas, or supplement existing ATC systems (53rd ATCS, 2020).[13]

Most of the Air Force's deployable ATC capabilities are in ten ANG ATCSs.[14] Their mission is to provide rapid deployment and to operate and maintain the ATCALS to ensure safe, efficient flying operations in both deployed and nondeployed environments. This "allows for safe reliable routes for U.S. military and allied nations' aircraft, utilizing the latest in tower and radar services" (ANG, undated).

Regional Maintenance Centers

The Air Force manages maintenance of ATCALS regionally through three RMCs. The RMC in Oklahoma City, Okla., services CONUS, Greenland, and South and Central America. The RMC located at Kapaun Air Station, Germany, services sites in Europe and Africa. The RMC located at Yokota AB, Japan, services sites in the Pacific, including Alaska and Hawaii (AFM 13-204, Vol. 4, 2020b).

In addition to providing on-site maintenance for equipment within their respective regions, RMCs may also be called upon to support other regions. The RMCs provide traditional Air Force Engineering and Technical Services and ATCALS radar evaluation support for all units.

With this summary of the AO career fields' supply, we now consider the demands that are placed on personnel in these career fields.

---

[12] The 821st CRG at Travis Air Force Base (AFB), Calif. (organizationally under the 621st Contingency Response Wing [CRW]), the 621st CRG at Joint Base McGuire-Dix-Lakehurst (JBMDL; also organizationally under the 621st CRW), the 435th CRG at Ramstein Air Base (AB), Germany, and the 36th CRG at Andersen AB, Guam). CRSs are the 36th CRS at Andersen AB; the 321st CRS and the 621st CRS at JBMDL; the 435th CRS at Ramstein AB; and the 821st CRS and 921st CRS at Travis AFB. The 123rd CRG (ANG) is at Kentucky ANG Base, Louisville, Ky.

Today these units have a mobility aircraft focus and provide only limited runway operations after the arrival of the main force (Schmidt, 2020). They also have limited capabilities with respect to landing zone control, landing zone safety officers, and drop zone control. While the Air Force is piloting efforts to make CRGs "aircraft agnostic" and to enable "full spectrum air base operations" (Schmidt, 2020), existing CRGs appear to be undermanned. RAND's database shows a total of 33 unfilled authorized CRG positions at Anderson AB, JBMDL McGuire, Ramstein AB, and Travis AFB.

[13] Like the CRGs, the 53rd ATCS also appears to be undermanned: There are 129 personnel assigned to the 53rd, but RAND's database shows that there are 34 unfilled authorized enlisted positions. Undermanning in these areas may mean that the Air Force needs to reconsider *where* its AO resources should be located (OSSs or specialized units) to be available for deployment in the future operating environment.

[14] Interview with National Guard Bureau, May 3, 2021.

Employment of Airfield Operations Personnel

Categorizing Employment

AO personnel can be stationed in-garrison or deployed, and the nation can be in one of two states of what joint doctrine publications call *the competition continuum*: in competition or in a contingency.[15] Thus, AO capabilities can be viewed from the perspective of the four states illustrated in Table 2.3. Personnel in-garrison while in the *in competition* range of the competition continuum have a variety ATC, airfield management, and equipment responsibilities. These can support AO and unit training activities, unit exercises, and daily flying operations.

Table 2.3. Categorizing AO Employment

Mission Location	In Competition	Contingency State
In garrison	• AO capability/unit training • Supporting training • Exercises • Daily operations	In competition missions *plus* • Kinetic missions, such as launching combat operations from CONUS locations • Engagement in other theaters, such as for humanitarian assistance and disaster relief (HADR)
Deployed	• AO capability/unit training • Supporting training • Exercises • Daily operations *Different skills may be emphasized because of deployed infrastructure or interaction with host nations*	In competition missions *plus* • Kinetic missions, such as conducting combat operations from forward operating bases and/or austere airfields • HADR

[15] U.S. Joint Chiefs of Staff, 2017, incorporating change 1, October 22, 2018, describes a "conflict continuum" that ranges from peace to war (p. i-4). A later document, Joint Doctrine Note 1-19 (O'Donohue, 2019), refers to a "competition continuum." Rather than a world either at peace or at war, the *competition continuum* describes a world of enduring competition conducted through a mixture of cooperation, competition below armed conflict, and armed conflict (p. v):

> Cooperation and competition below armed conflict are always occurring and so the presence or absence of armed conflict is the only variable element. Therefore, the notions of the US being either "in competition" or "in conflict" actually refer to whether the joint force is or is not using armed force as a means to achieve policy objectives. In either case, the joint force will be conducting cooperative activities with partners and competitive activities below armed conflict to counter adversaries who are seeking to turn the competition or conflict to their advantage. The joint force is never solely in cooperation (or in competition below armed conflict or in armed conflict) but instead campaigns through a mixture of cooperation, competition below armed conflict, and armed conflict calculated to achieve the desired strategic objectives.

We use *in competition* and *contingency state* rather than *peace* and *war* to be consistent with the later publication.

The *contingency* state indicates that some form of contingency order has been executed by various levels of command authority up to the President of the United States.[16] Contingency operations are not limited to near-peer war operations and can include less-kinetic activities like "peacekeeping operations" and HADR. While in the contingency state of the conflict continuum, personnel in-garrison will still be responsible for their in-competition duties but may have increased demands placed on them if, for example, combat operations are launched from CONUS. For our purposes, the contingency state is one where an airfield is conducting other than routine day-to-day activities under the direction of higher headquarters, either for exercises or force generation and deployment. An example would be if an installation is tasked to deploy a portion of its force, and the airfield may still have to remain open beyond published hours to accommodate transportation operations.

Personnel deployed while in the *in competition* range of the conflict continuum perform duties that are no different from those they do at home. However, deploying may require an emphasis on different skill sets. This part of the matrix accounts for current deployments, such as those required to support U.S. Central Command (USCENTCOM) operations.

Personnel deployed while in the contingency range of the conflict continuum might face the challenge of performing their duties from FOBs or austere locations.

Quantifying Demand

To help understand the potential constraints on the availability of AO personnel for new missions, it is useful to understand the difficulty associated with quantifying demands for these personnel.

In-competition Demands

The number of people authorized to be assigned to AO units is determined by established manpower standards (U.S. Air Force, 2007) and is derived from calculated workloads, which are in turn based on hours of operation. The 2007 manpower standard states that these numbers apply to home station operations during peacetime and wartime. But two key concerns about the existing manpower standards are that they are out of date and that they do not account for deployment demand.

Current manpower standards do not reflect changes in the AO career fields, such as the merger of several career fields to create the RAWS career, as previously discussed. A study to update the manpower standards began in 2017 and is expected to conclude in 2022. However,

---

[16] Air Force documents define a *contingency* as "a situation requiring military operations in response to natural disasters, terrorists, subversives, or as otherwise directed by appropriate authority to protect United States interests" (AFMAN 13-204, Vol. 1, 2020a). Air Force tactics, techniques, and procedures documents describe how to open and operate a deployed or contingency airfield, planning considerations for airfield management, ATC, and deployable ATC and landing systems capabilities, procedures, forces, and equipment (423rd Mobility Training Squadron, 2017).

while the study includes requirements for AO staff, airfield management, and ATC, it excludes RAWS because the career field was deemed too immature for analysis in 2017 (Mineau, 2021, p. 3).

More importantly, the manpower standard does not account for deployment demand. Unit sizes are based on a calculated peacetime workload, so if personnel are deployed, those who are left behind must accomplish the same workload with fewer people—unless operating hours are changed or other services are reduced.[17]

After 20 years of deployments in the USCENTCOM theater of operations (Operation Enduring Freedom began in 2001), enduring deployment commitments are well understood,[18] and those requirements plus those of U.S. Africa Command, U.S. Northern Command, and U.S. European Command totaled about 375 billets in 2021.[19] According to interviewees, the AO community has, over the years, used what they called a crowdsourcing approach to support USCENTCOM needs. This approach involved pulling individual personnel from various locations to fill a set of USCENTCOM requirements rather than attempting to deploy people who may have worked together at their home unit. In addition, some of these deployment demands are "hidden," in the sense that, over the years, USCENTCOM has come to depend on contractors to fill from 175 to 200 AO billets (as noted previously).

Contingency Demands

AO in a contingency environment is described in Air Force Tactics, Techniques, and Procedures Vol. 3-4.4, *Contingency Airfield Operations*, 2017, which defines needed AO functions based on organization, capability, and employment in a given contingency environment.

Personnel requirements for these functions will be defined in operational planning documents, but additional demands may also be placed on the AO community. For example, during hurricane relief operations after hurricane Maria in 2017, ANG units provided two mobile towers to Puerto Rico.[20] As a participant in the FAA Contract Tower (FCT) program, the ANG also has an agreement with the FAA for six airfields that, in the event that ANG air traffic

---

[17] According to Robbert et al., 2020, AFMAN 38-208, Vol. I, 2002, pp. 107–110, provided computation methods for deployment credit, but the manual that replaced it in 2019 (AFMAN 38-102, 2019) does not mention deployment credit.

[18] Interview with personnel from USCENTCOM, January 9, 2021.

[19] According to unclassified numbers provided by U.S. Air Force Personnel Center (AFPC)/Directorate of Air and Space Expeditionary Force Operations).

[20] Interview with personnel from the ANG, May 3, 2021.

controllers are deployed, the Guard will ensure that civilian air traffic controllers will maintain tower operations at those locations.[21]

An active-duty CONUS location, such as a bomber wing, could find that it is shorthanded because it has deployed AO personnel for contingency operations. However, since bomber combat missions can be launched from CONUS locations, this shorthanded unit could also find itself responsible for additional services related to combat missions for the same contingency.

As another example, the Emergency Security Control of Air Traffic (ESCAT) preparedness plan prescribes the joint action to be taken by appropriate elements of DoD, the Department of Transportation, and the Department of Homeland Security in the interests of national security to control air traffic under emergency conditions (FAA, 2018, p. 2).[22] SMEs we interviewed at Air Force Northern Command (AFNORTH) indicated that the FAA would not depend on military personnel to assist with ATC responsibilities in these conditions.[23] However, AFNORTH personnel noted that 20 CONUS locations have regular air defense responsibilities.[24] The need to ensure the maintenance of ATC or airfield management functions associated with air defense could constrain the availability of AO personnel for deployment at those locations.

Finally, during contingency operations, additional AO personnel may be needed at fixed CONUS or OCONUS locations to support global mobility *air bridge* requirements related to moving personnel and equipment to the area of contingency operations. Table 2.4 summarizes the characteristics of demand for AO personnel in the states described in our matrix—some of which could constrain the supply of personnel for other deployments—and highlights where these demands are ill-defined. A full understanding of these demands and constraints is necessary before the Air Force can quantify the availability of AO personnel for increasing deployment demands.

---

[21] As stated by the FAA, "To ensure continuity of operations, the [National Guard Bureau] shall staff a minimum of five (5) GS 2152 controllers or Military equivalent (AFSC 1C1XX/13MX) at each FCT." FCTs represent 49 percent of all federal ATC towers in the United States (FAA, 2020; and interview with personnel from the ANG, May 3, 2021).

[22] Emergency security control of air traffic procedures are noted as a specialty-wide core task in the Career Field Education and Training Plan for the 1C1x1 (ATC) career field.

[23] Interview with personnel at AFNORTH, June 29, 2021.

[24] The 20 regularly planned locations include 14 ANG fighter alert sites, four tanker sites, one airborne early warning (AEW) site, and one Coast Guard site. An additional ten locations can be added if necessary (AFNORTH interview, June 29, 2021).

Table 2.4. AO Demands That May Be Ill-Defined

Mission Location	In Competition	Contingency State
In garrison	• Defined by manpower standard • Potential national airspace demands from agreements with FAA	• No formal demand numbers for homeland security; demand for HADR operations will vary • Personnel to maintain ATC and airfield management needs of locations responsible for air defense • ANG agreements with FAA to use civilian personnel to maintain AO
Deployed	• Approximately 375 enduring deployment requirements for U.S. Africa Command, USCENTCOM, U.S. European Command, and U.S. Northern Command • "Hidden" demand: 175–200 AO billets filled by contractors	• U.S. Transportation Command air bridge requirements for AO personnel for Global Mobility Laydown requirements • Operations plan requirements

SOURCE: Enduring deployment data from AFPC.

Airfield Operations Equipment

AO personnel require several types of specialized ATCALS and deployable ATCALS to perform their duties:

- training (ATC tower simulators)
- navigation (fixed or deployable Very High Frequency Omnidirectional Range [VOR] systems, Very High Frequency Omnidirectional Radio Range/Tactical Air Navigation [VORTAC] systems, and Tactical Air Navigation Systems [TACANs])[25]
- landing guidance (precision approach radar and instrument landing systems)
- radar approach control systems
- communication (ATC radios)
- airfield management vehicles or trailers

Table 2.5 lists types of fixed and deployable AO equipment currently in the USAF inventory, from oldest initial operating capability to newest. The age of some of this equipment is an important limitation on AO capabilities, as will be discussed in later chapters of this report.

[25] VOR and TACAN are radio transponder navigation systems. A VORTAC is a collocated VOR and TACAN.

Table 2.5. Airfield Operations Equipment

Equipment	Initial Operating Capability
Deployable radar approach control	1960s–1970
Fixed-base TACAN/VOR/VORTAC	1960s–2015
Fixed-base precision approach radar	1970s
Fixed-base instrument landing system	1970s–2013
ATC radios	1970s–2016
Man-portable TACANs	1978
Deployable ATC towers	1999
ATC tower simulators	2002
Airfield management vehicles/trailers	2002
Deployable TACAN	2005

SOURCE: Personal communication from Headquarters Air Force Flight Standards Agency (AFFSA) Program and Requirements Directorate.

The Air Force has assigned a significant portion of its deployable AO equipment to the ten Air National Guard ATC squadrons: 59 percent of its mobile towers, 71 percent of the deployable Radar Approach Control systems, and 50 percent of the deployable TACANs (Roberts, 2021). The radars in the deployable ANG "suites" (designated MPN-14K) are different from the TPN-19 radars used in the deployable packages of the active-duty 53rd ATC Squadron (National Guard Bureau, 2021).

According to interviews with SMEs, a potential issue for deployment is that the two radar systems require different training, and both have steep learning curves, so ANG personnel could not deploy to a location that uses active-duty equipment and vice versa unless additional training were provided.

The size of the AO career field, the components where personnel reside, and the services they perform in both the *in competition* and *in contingency* states determine the availability of personnel for contingency operations. Understanding this relationship is important as the Air Force evolves its concepts of operations (CONOPs) to meet a new deployment strategy, discussed in the following chapter.

3. DFE and ACE Concepts and Their Potential Impact on Airfield Operations

In the past three years, there have been changes in national security guidance, operational concepts, and how the USAF postures its forces. All three affect the supply of, and demand for, AO personnel and equipment. Maneuver and more-dispersed basing mean more operating locations, which increases the demands on AO personnel and equipment. This chapter describes current guidance, the ways in which the USAF is responding to that guidance, and how those changes affect the AO community.

The 2017 National Security Strategy and 2018 National Defense Strategy

Both the National Security Strategy (NSS) of 2017 and the National Defense Strategy (NDS) of 2018 discuss the increasingly complex security environment, which poses challenges to the U.S. military advantage (Trump, 2017; Mattis, 2018). The NSS describes four national interests,[26] and the NDS describes the defense objectives and approach to achieve them. Protecting American people and defending the homeland are the first priorities in both strategies, and both acknowledge that adversaries are challenging the United States on economic, political, technological, and military fronts.[27]

The United States has operated from relatively secure, safe-haven locations for many years; however, current near-peer competition has the United States focused on being ready to protect the nation and deter adversarial kinetic and nonkinetic attacks from contested and degraded operational environments. The NDS emphasizes "being strategically predicable, but operationally unpredictable" and recognizes that one way to achieve that goal is by developing "a lethal, agile, and resilient force posture and employment" utilizing DFE (Mattis, 2018).

DFE, as defined in the NDS, consists of scalable joint forces able to scale up to conduct major combat operations and scale down to provide proactive, rapid-response deterrence capabilities. This flexibility allows the joint force to be more agile and maintain its combat readiness, all on a schedule that is unpredictable to the adversary. The key tenets of DFE are flexibility, unpredictability, and proactive operations.

---

[26] The four national interests are (1) protect the American people, the homeland, and the American way of life; (2) promote American prosperity; (3) preserve peace through strength; and (4) advance American influence (Trump, 2017, p. 4).

[27] The *Interim National Security Strategic Guidance*, published in March 2021, emphasizes restoring and reinvigorating U.S. national strength (Biden, 2021).

The DFE force management framework[28] bins forces into three categories according to readiness and availability to infuse more predictability into U.S. and ally understanding of global force movements (Esper, 2020):

- Immediate response force: Forces in this category are the most ready and available for surprise deployments to enhance operational unpredictability. Immediate response forces also provide rapid contingency response and serve as a "blunt" force in conflict.[29]
- Contingency response force: This force is less ready than the immediate response force but still able to surge to support a blunt force.
- Force generation forces: Units in the force generation forces bin are not ready or available to deploy because they are either building readiness at home or are already deployed.

DFE in the U.S. Air Force

The USAF has exercised the DFE concepts of unpredictability and flexibility in several ways, two of which we discuss in this section: bomber task forces and the CONOP the USAF is calling ACE.

The U.S. Global Strike Command regularly executes Bomber Task Force (BTF) missions to demonstrate flexibility while training with partner nations and familiarizing personnel with different geographic areas of responsibility (AORs). The goal of the BTF is to deploy to different geographic locations, operate for a period (not on a predictable schedule), and exercise different capabilities as a way to demonstrate the USAF global strike force's ability to meet mission requirements.

The ability of the BTF to go anywhere at any time is intended to serve as a deterrent option. For example, in March 2021, B-2 bomber aircraft from Whiteman AFB, Missouri, and B-1 bombers from Dyess AFB, Texas, flew to Lajes Field, Azores, to train with the Portuguese military and operate within the European theater. The B-1 conducted integrated operations with allies in Norway and performed refueling operations in Poland before returning to Texas (Salazar, 2021). BTF operations to demonstrate U.S. capabilities are ongoing in support of several geographic combatant commands (COCOMs).

The ultimate goal is for U.S. forces to be able to execute the mission close to the point of need, including operating in contested and degraded operational environments. How those forces

---

[28] As DFE evolves, there is some discussion about whether it is a *strategy* or a *framework*. The Joint Staff characterizes it as a *framework* (Joint Staff, 2018).

[29] The 2018 NSS associates a Global Operating Model with DFE. This model comprises four layers (see Mattis, 2018):

- Contact: designed to help the United States compete more effectively below the level of armed conflict
- Blunt: delay, degrade, or deny adversary aggression
- Surge: surge war-winning forces and manage conflict escalation
- Homeland: defend the U.S. homeland.

plan to operate depends on the CONOP for that AOR. Across AORs, emerging CONOPs and concepts of employment (CONEMPs) are incorporating some form of maneuver. By *maneuver* we mean moving the geographic location of air operations and the ground support for those air operations in an unpredictable manner. How maneuver is defined varies in frequency and timelines across AORs.

PACAF has outlined an ACE CONEMP defining how PACAF plans to deter or, if necessary, defeat a near-peer adversary in a contested environment. *ACE*, as defined by PACAF, is "a proactive and reactive operational scheme of maneuver executed within threat timelines to increase survivability while generating combat power." ACE operating concepts require "more responsive, resilient, and agile" forces and build on current CONEMPs, such as cluster operations, in which distributed hub-and-spoke locations operate in a network to "maintain operational momentum in a high-end fight, defend allies and partners, and present layers of operational unpredictability to complicate an adversary's strategic calculus" (Department of the Air Force, PACAF, 2020). The tenets of PACAF's ACE CONEMP are directly linked to the 2018 NDS and DFE.

Both DFE concepts discussed in this section, the BTF and ACE CONOPs, involve moving aircraft to different locations. Because not all the aircraft may relocate or maneuver (some aircraft may remain operating from home station), there could be an increase in the number of locations that require AO personnel and equipment. Unless home station locations are closed, home station service is reduced, or personnel are sourced from other locations, AO personnel could be trying to operate multiple locations out of the manpower authorized to support a single home station location.

Notional ACE Cluster Components

In this section, we define a notional operational construct and force laydown that could be used to support an ACE scheme of maneuver. Using PACAF terminology, we define a *cluster* as a single *hub* plus four *spokes*, each with differing operational capabilities as described in Table 3.1. *Drop-in* locations may be established to support one or more clusters, depending on their geographic location. Each of these cluster components will have *mission generation* capabilities, meaning that aircraft will be taking off and landing from that location; therefore, each of these components with require some level of AO capabilities to enable that sortie generation capability.

Table 3.1. A Notional Cluster, Its Components and Capabilities

Cluster Component	Operational Definition	Mission Generation and BOS-I Capability
Home station	Steady-state assignment	Full capability
Hub	Main forward deployed operating location	Full deployed capability
Spoke	Secondary forward deployed operating location, associated with a single hub	Limited capability
Drop-in location	Further forward (preferable) deployed operating location, for resupply during a mission	Extremely limited capability (for example, refuel and rearm only)

SOURCE: Adapted from Lynch et al., 2022, Table 2.1.
NOTE: BOS-I = base operating support–integrator.

USAFE is also developing and exercising ACE concepts where forces can operate from different locations with those locations providing varying levels of operational support capability. USAFE too is using a hub-and-spoke construct, although they use different terminology, such as *main operating base* (MOB), FOB, and FOS.[30] As in the PACAF AOR, each of the operating locations used in the USAFE scheme of maneuver will require some level of AO capability.

ACE Deployment and Beddown

We define deployment to a notional cluster construct as a two-step process (Figure 3.1). First, we assume that squadrons deploy aircraft and associated unit-owned capabilities from their home stations to the cluster hub or MOB. This leaves some aircraft, residual unit capability, and other home station capabilities back at the home station location. Then, squadrons disperse to a spoke or FOB, with some level of capability, leaving one squadron at the hub. Again, some residual capability may remain at the hub or MOB, depending on the level of capability that the unit takes with it to the spoke or FOB.

AO resources could be divided to support ACE-like operations. Some AO resources should be deployed from home station to the hub to support the hub flying operations, and some residual AO capabilities should remain at home station to continue the flying operations in garrison. If aircraft maneuver from the hub to multiple spokes, the AO capabilities are further divided to support all spoke locations (unless AO personnel and equipment are supplemented from other locations). As discussed previously, increasing the number of operating locations increases the demand on AO capabilities.

---

[30] As a matter of historical interest, in 1954, USAFE proposed a dispersed basing concept to protect assets from attack by the USSR during the Cold War. From an MOB, resources could be distributed to dispersed operating bases to conduct operations, dispersed landing areas to recover aircraft, and dispersed parking areas. See Benson, 1981 (declassified 2010).

Figure 3.1. Deployment to a Cluster Is a Two-Step Process

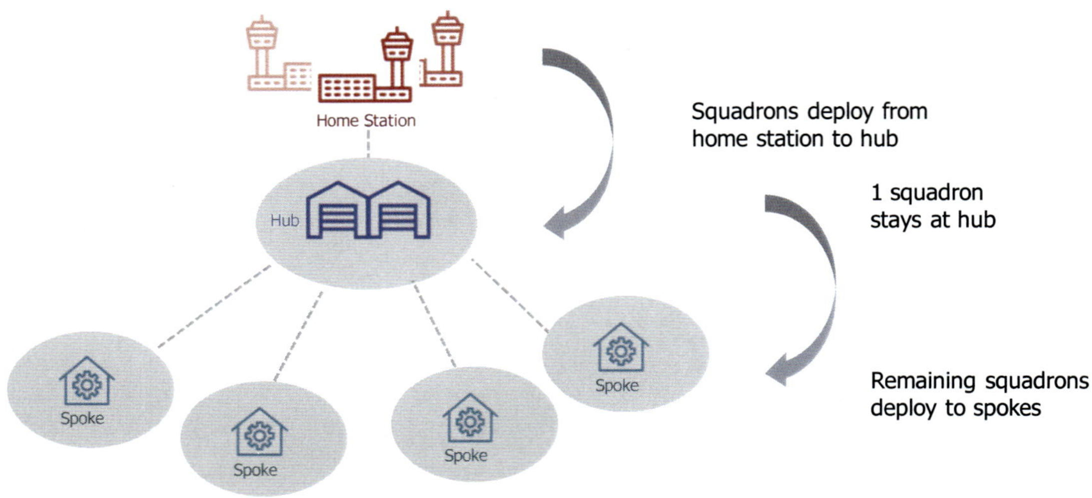

SOURCE: Lynch et al, 2022.

These increasing demands become a complex issue for AO. How much AO capability remains behind at the home station location to maintain training operations there? How much AO capability is required at the spoke or FOS? If AO capability moves forward to the spoke or FOS, how much capability remains at the hub or MOB to continue operations at that location? These questions and others will be evaluated and assessed in Chapter 5, when we compare the demand for AO capabilities against the supply of available AO resources.

Multi-Capable Airmen

One force posture concept currently being developed across the USAF to support ACE-like concepts is teams of multi-capable airmen (MCA):

> Multi-capable Airmen accomplish tasks outside their core Air Force Specialty, specifically trained as a cross-functional team to meet and provide combat support to aviation force elements conducting dispersed and dynamic operations. Enabled by cross-utilization training (CUT), they operate independently to accomplish mission objectives within acceptable levels of risk. (Department of the Air Force, Pacific Air Forces, 2020).

The U.S. Air Force has some experience with increasing the number of tasks individual airmen are expected to accomplish: It initiated a multiskilling effort in the 1980s known as Project Rivet Workforce as part of the general drawdown in force structure (both aircraft and people) following the end of the Cold War.[31] This initiative combined maintenance AFSCs based on similar technologies and infrequent usage.

A recent RAND analysis evaluated the costs and benefits of consolidating maintenance career fields for the F-35A in terms of managing the force and meeting emerging operational requirements. The analysis found, among other conclusions, that merged AFSCs contribute to readiness to support ACE-like concepts because of the smaller footprints needed to support a forward location; however, it would take many years to institute the training necessary for maintenance personnel to obtain skills across multiple specialties (Wirth et al, 2020).[32]

Instead of combining AFSCs, the most recent version of the PACAF CONEMP delineates tasks in each career field and then identifies other personnel who may be able to complete those tasks. Not every task can be completed by personnel from another career field. For example, weapons loaders cannot provide ATC—personnel in each of those career fields have to be certified to do those jobs. Other tasks, such as setting up expeditionary tents, may be accomplished by any USAF personnel without in-depth, content-specific training.

The PACAF CONEMP provides tiered levels of MCA support. The tiers range from expeditionary tasks that can be accomplished by all airmen (like setting up a tent, driving a vehicle, self-protection fundamentals, and cargo preparation) to PACAF-specific expeditionary and mission generation tasks that may pose risk if they are completed by others without more training (for example, armed perimeter security, driving heavy machinery, or repairing air traffic and warning radar systems).

Airmen have been participating in MCA events, across the USAF and across AORs, since the Air Force Installation and Mission Support Center began its Combat Support Wing experiments with MCA in 2015. The 18th Wing at Kadena AB, Okinawa, Japan, hosted its first MCA course in February 2021. Part classroom education and part hands-on training, the MCA course at Kadena was designed to give airmen understanding (and experience) of capabilities required for ACE-like operations beyond their own AFSC (Nolte, 2021).

There are several challenges associated with the MCA concept. First, training—beyond what is required for initial proficiency—is needed to help airmen retain additional skillsets. That training is an additional burden to units. Second, there is risk associated with the MCA concept. What the USAF gains in breadth, it could lose in depth. Becoming a generalist rather than a specialist in a high-end fight may pose a risk. For example, aircraft battle damage repair and fifth-generation aircraft maintenance may require highly skilled airmen with deep knowledge to

---

[31] See Reiter, 1988. In 1987, during the time of Rivet Workforce, the maintenance manpower was at a high point of about 172,000. Today, there are about 84,000. Similarly, the number of aircraft in the fleet in 1987 was about 9,450 and today is about 5,350 aircraft.

[32] According to a November 19, 2021, e-mail from the author of the F-35 RAND analysis, the USAF is awaiting more data on F-35 maintainability and reliability to determine the optimal organization of F-35 maintenance career fields. RAND (DeBlois et al., 2020) has also studied the possibility of combining AFSCs in relation to KC-15 maintenance.

repair aircraft to get them back into the fight. Similarly, an air traffic controller requires certifications and proficiency on each location that they control. The ATC training and experience requirements might make that position one that other personnel could not easily train for. Airfield management, on the other hand, may be an AFSC that other personnel could be trained for, along with their primary AFSC.

There is also risk in aligning too many capabilities with too few people. Forward locations (spokes, FOBs, and FOSs) are intended to be small and employ MCA. If there is personnel attrition in a high-end fight and an MCA is lost, the unit loses all the capabilities for which that individual was responsible. There is a risk that a forward location could lose too much capability by losing a few personnel.

If the USAF were to cross-train personnel from other career fields to perform AO duties, it would relieve some of the burden on the small AO career field. There would be more personnel trained to provide AO capabilities to support increasing requirements, such as the hub-and-spoke concepts previously described.

The Air Force Force Generation Model

Since the 1990s, the USAF has employed an organizing construct to distribute deployable resources (personnel and equipment) into bins of capability that may be used on a rotational basis to meet enduring and emerging operational requirements. After Operation Desert Storm in 1991, operations continued in Southwest Asia for years in support of Operation Southern Watch and Operation Northern Watch. The continued deployment and the need to act quickly led the USAF to develop the air and space (or aerospace) expeditionary force (AEF) construct to manage its resources. The AEF construct divided USAF capabilities into ten roughly equivalent "buckets." The ten buckets were paired and divided into five rotational cycles (of approximately three months) in which the resources within those buckets would be eligible for deployment within that three-month cycle. The AEF Center, located at that time at Langley AFB, Virginia, managed the AEF buckets and scheduled the deployments.[33]

Over the years, the AEF construct evolved into more of a scheduling tool than an operational force provider tool. Personnel in the deployable buckets were used to fill individual requirements rather than being deployed as part of a capability. In 2013, then Chief of Staff General Mark Welsh championed an AEF Next initiative that would return the AEF construct to providing units, teams, or capabilities for deployment, instead of individual personnel (Brown, 2014). Although approved, the AEF Next construct was never implemented.

With enduring deployed operations continuing and the recent focus on military readiness, the USAF is developing a four-phase AFFORGEN model (shown in Figure 3.2). The USAF force

---

[33] In 2006, the AEF Center became a direct reporting unit of the AFPC, and in 2007, it was merged into the AFPC as a separate directorate and relocated to Joint Base San Antonio, Texas. AFPC is a field operating agency of Headquarters, U.S. Air Force, Office of the Deputy Chief of Staff, Manpower and Personnel. See USAF, 2015.

presentation model covers a 24-month cycle, with six months per phase. The intent is to focus on readiness while identifying what USAF capabilities are available to fight tonight. The AFFORGEN model, like the original AEF model of the 1990s, is a construct to provide ready forces to the warfighter when needed.

Figure 3.2. AFFORGEN Model

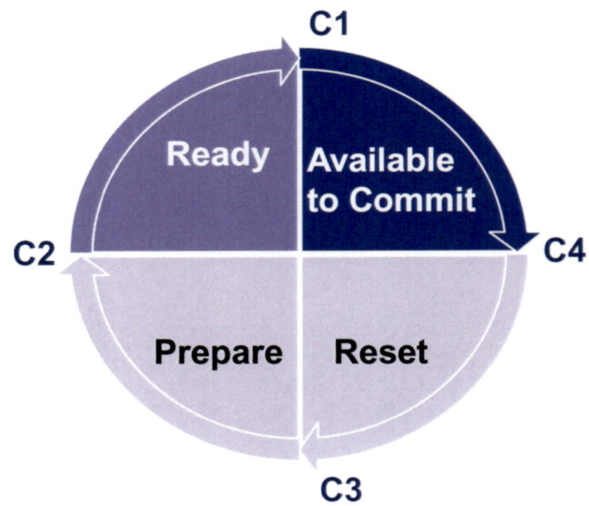

SOURCE: U.S. Air Force, 2021b.

Forces in the AFFORGEN model will be presented in force modules. These force modules consist of both combat and combat support capabilities that are grouped together to form scalable, modular force elements. Six of the seven force elements are shown in Figure 3.3.[34]

Figure 3.3. AFFORGEN Force Elements

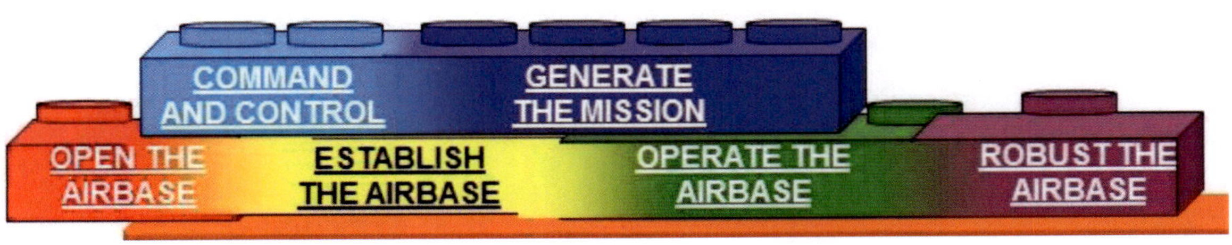

SOURCE: USAF, 2015.

The AO mission is represented in several of these force elements—open the air base, establish the air base, operate the air base, and generate the mission. A challenge for the AO

[34] Figure 3.3 does not show the "close the air base" force element.

career field will be to determine where it best fits into the AFFORGEN force presentation construct.

The AFFORGEN force *generation* and DFE force *management* models can be related, as shown in Figure 3.4. Air Force forces in the Reset and Prepare phases would be considered part of the force generation forces—building readiness. Forces in the Ready phase are available as Contingency Response Forces, and those in the (ready to) Commit phase would be considered part of the immediate response forces. With the 24-month cycle and six months per phase mentioned above, the Air Force would maintain a 1:3 deploy-to-dwell time (for each six months of operational deployment, an individual will then have 18 months of nondeployment) (Wright, 2013).

The understanding of AO personnel and equipment presented in Chapter 2 and the changing military environment discussed in this chapter provide context for understanding the existing "gaps" in AO capabilities and how the changing environment may exacerbate them, as will be discussed in the next chapter.

Figure 3.4. Associating DFE and AFFORGEN Concepts

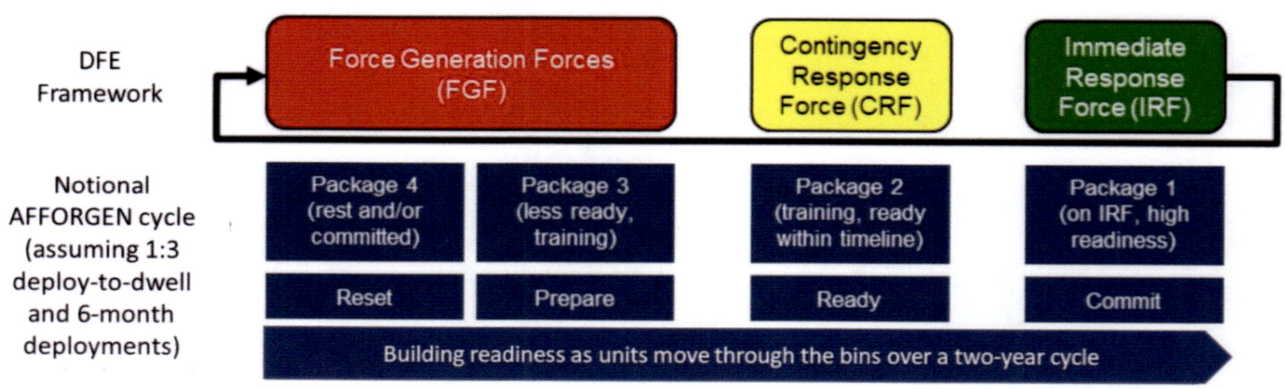

4. Airfield Operations Challenges

The introduction of DFE and ACE concepts has the potential to increase demands on the AO community: If the USAF must be prepared to establish and maintain more airfields as part of a dispersed installation approach, the supply of AO personnel could be stretched thin. To better understand the potential challenges that the AO community will face in a new strategic environment, we interviewed SMEs and summarized insights from the Vision and Bridging Strategy sessions that members of our team attended. This chapter describes challenges that were raised most often in these venues.

Challenges Assessed in a DOTmLPF-P Framework

After various interviews, discussions, and meetings with relevant stakeholders in the AO community, we identified ten capability challenges that could affect the AO community's ability to respond to emerging missions.[35] These challenges fall into four major categories: (1) manpower and personnel, (2) force presentation, (3) force generation, and (4) culture, as shown in Table 4.1.[36]

Table 4.1. Summary of Challenges

Challenge Category	Challenges
1. Manpower and personnel	a. Insufficient manpower; too-small career fields; crowdsourcing b. Demands on training, accession, retention, deployability, and mission success c. The role of the guard, reserve, civilians, contractors, and joint, interagency, intergovernmental, and multinational partners
2. Force presentation	a. Posture b. Antiquated equipment and infrastructure c. Impact of ACE and DFE/hub-and-spoke environment
3. Force generation	a. Accessibility and training b. Surge capacity and back-filling positions c. Requirements
4. Culture, pace of policy change, and community alignment	

[35] Appendix B contains details of the analysis we conducted to arrive at these capability challenges.

[36] Force generation is the process to provide a systematic, service-wide approach to selecting, resourcing, and preparing units for operational employment. Force presentation is the "preferred organizational construct through which a service offers and attaches its operational capabilities to the combatant commander for operational employment." See USAF, 2021a.

After identifying these challenges, at the request of the sponsor, we captured the frequency with which they were mentioned in interviews within a modified–DOTmLPF-P framework (which includes the addition of "mission management" and "strategic communication and coordination" categories), shown in Table 4.2 (AFI 10-601, 2013).[37] The cells in the table show the number of times that a topic was mentioned in interviews, and the color coding of cells indicates the number of venues where the topic was raised (see Appendix A).

Table 4.2. Capability Challenges Across Modified DOTmLPF-P++ Framework

	D	O	T	M	L	P	F	P	+ Mission	+ Strat C&C
Manpower & Personnel										
Insufficient manpower; too-small career fields; crowdsourcing	1	12	4	3	1	17			10	1
Training, accession, retention, deployability, and mission manpower demand	1	15	9	1	1	16			14	1
The role of ANG, Reserve, civilians, and contractors	1	18	4	3	1	17	1	1	13	8
Force Presentation										
Posture	7	36	11	11	5	17	1	3	17	9
Antiquated/heavy equipment and failing infrastructures	5	14	7	29	4	7	5	4	13	10
Impact of ACE and DFE; hub-and-spoke environment	5	22	8	7	2	9	7		14	4
Force Generation										
Accessibility and training	3	23	25	14	3	18	4	3	16	6
Surge capacity and backfilling positions	1	15	2	2		15	2	2	9	11
Requirements	7	26	5	5	1	17	1	7	19	7
Culture; Pace of policy change; Community alignment	17	14	1	2	2	3		12	16	10
[dark shading]	Number of strong implications (e.g., ten or more)									
[light shading]	Number of moderate implications (e.g., less than ten)									

NOTES: Strat C&C = strategic communication and coordination. The frequency counts in the table are aggregations of frequencies within each data stream (i.e., interviews, strategy-vision bridging working sessions, and the executive council presentations). See Appendix A for a discussion of how the distinction between strong and moderate implications was determined.

For example, as shown under *force presentation*, challenges related to AO posture exist within the organization (O), training (T), materiel (M), personnel (P), and mission management areas (+ *mission* in the table). Challenges related to culture, community alignment, and the pace of policy change versus the pace of emerging missions primarily manifest across doctrine (D), organization (O), personnel (P), mission management (+ mission), and strategic communication

---

[37] In our analysis of interview comments, we found that two additional operational categories were useful: *mission management* and *strategic communication and coordination*; these are the "++" categories in our discussion.

and coordination (+ Strat C&C). In the aggregate, the two management areas with the most challenges are organization and mission.

While materiel did not have the most frequency hits, it is important to note that equipment-related concerns emerged within multiple challenge categories (e.g., impact of manpower demands on training, the role of ANG, the impact of ACE and DFE, accessibility and training, and requirements). Moreover, equipment-related concerns were highlighted in all but one interview, as well as the strategy-vision bridging sessions and executive council presentations. In the following sections, we primarily address equipment challenges in the force presentation category of concern, but equipment-related issues also emerge throughout each section.

We discuss the challenges in the following sections by category.

Manpower and Personnel Capability Challenges

Three issue areas emerged within the manpower and personnel–related category:

- the lack of manpower, too-small career fields, and problems created by crowdsourcing (i.e., selecting individuals from diverse units to fill deployment requirements)
- the impact of manpower demands on accession, training, retention, deployability, and mission success
- the role of the ANG, AFR, civilians, and contractors.

Impacts of Limited Manpower Supply and Increased Mission Demand

According to our interviews and insights gleaned from strategy-vision bridging sessions and the executive council briefings, there are significant concerns relating to the size of the AO community, particularly the 13M (AO officers) career field. As some of our interviewees stated,

> [Our] biggest problem is we are too small. 13M community on core authorizations are right around 240 [from 2Lt through Lt Col]. We are in the bottom 15 specialties in the Air Force. Of all traditional Air Force functional communities, an argument could be made that 13M is the smallest.
>
> The [career field managers] don't have enough people to do the stuff they are doing.

Manpower shortages within the AO community can have significant repercussions at home both in peacetime and in a contingency. In peacetime, the noncommissioned officer shortage translates to reduced training capability, which, in turn, manifests in an inability to meet wartime demands. Moreover, bases are manned to support only home station missions, which creates risk when downrange mission (deployment) requirements emerge. Interviewees told us that Air Force documents no longer define *critical* manning levels, but that there are manning levels below which units cannot go while still maintaining their missions.[38] These levels vary by location and

[38] In a December 15, 2020, interview with the AFFSA, we were told that AFIs used to specify that if a unit's manning was under 75 percent, the unit was below *critical staffing*, but that this definition no longer exists. Note

mission, but planners must be aware of the effect of low manning levels as they develop approaches to the ACE hub-and-spoke model and assign personnel and units to the four bins of the AFFORGEN force presentation model. In addition, crowdsourcing is frequently used to pull individual personnel from across active-duty locations to support deployed missions, which can increase the time for people to learn how to work together as a team.

Manpower and personnel–related concerns were mentioned in 14 out of 19 interviews, as well as six strategy-vision bridging sessions. For example, one interviewee noted that "we found shortly after 9/11 that units didn't have manpower to support home station ops" and deployments, which led to "crowdsourcing" of individuals from three to four locations to "level the load." Regarding the impact of manpower shortages on mission execution, interviewees stated that

> I don't see our mission getting smaller, I see it potentially growing and the demand for our expertise growing but no real growth in our field because we are considered non-rated ops. With current accession rates, our manpower resources are going to be the biggest challenge for our community going forward, both in garrison and in deployment.

As one interviewee stated, "there are plenty of people being brought into AO, but there is an issue keeping them in AO." The AO community directly competes with the FAA and the airline industry for personnel, and the Air Force often loses out to federal agencies and private industry in terms of pay, location, and other benefits.

Impact on Individual Training

Interviewees and strategy-vision session participants also discussed how small career fields and manpower shortages affect individual training, which is connected to deployability. For example, an interviewee mentioned that "when 13M gets deployed, we don't have enough airfield managers—it looks like we have people, and we don't." Also, the shortage of NCOs (grade E-5—staff sergeant—and above) affects the ability to train other personnel and their availability for deployment: Skill level 7 is required to be deployed, and a person cannot begin 7-level training until selected for staff sergeant. Other interviewees stated,

> We are not right-sized for our current organizational structure. . . . What we do in-garrison distracts us from training and working on competencies for downrange requirements.

Interviewees mentioned that while they are organized, trained, and equipped "to do our in-garrison mission in peacetime, we are not for our wartime mission—and until we are, we will always have the same problem of providing consistent manpower downrange."

---

that this is different from the term *critical personnel*, used in the Defense Readiness Reporting System–Strategic (DRRS-S), which applies to personnel whose absence "would render the unit severely deficient or totally unable to execute its mission." (See AFI 10-201, 2020.)

The perception that AO is overly focused on its peacetime mission feeds into the common narrative from our interviews that the community is not prepared for contingencies.

Determining the Role of Civilians, Contractors, and the Reserve Component

While one interviewee stated that "The aircraft do not care who is in the tower," concerns related to the role of civilians, contractors, the Guard, and reserve were represented in 17 out of 19 interviews, as well as six strategy-vision bridging sessions. With respect to civilians, interviewees stated that, while civilians have played a key role in the stability and continuity of the in-garrison mission, the number of civilian airfield managers can make it difficult for military AO personnel to obtain experience in those positions.

Contractors have filled positions in locations such as Afghanistan, enabled by "bottomless Overseas Contingency Operations money," but some felt that this hinted at problems in the career field: "If we can't support downrange, why do we exist?"[39] In addition, according to SMEs, contractors are expensive, and using them contributes to a lack of stability in the workforce.

With respect to the use of the ANG, interviewees stated that, for example,

> [There are] ten expeditionary control squadrons in the Guard; it's time for those air traffic control squadrons to grow into providing airfield operation services . . . if we are going to make sure the Guard looks, trains, and develops like our active force, [because] our active-duty force transitioned to airfield operation 25 years ago . . . but the Guard 13M primarily are only focused on their air traffic [control] mission.

According to interviewees, without a presidential approval for mobilization, it is challenging to use ANG resources because "they have to volunteer to fulfill a requirement and can only be active for 120 days to allow them to go back to their normal civilian jobs . . . and they don't have equipment [that works], which isn't their fault."

There is currently no deployable capability in the AFR; interviewees suggested that this reflected an unwillingness on the part of the USAF to provide the manpower (which would have to be balanced by reductions in active-duty manpower).

Force Presentation Capability Challenges

Three issue areas emerged within the force presentation category:

- posture
- equipment and systems
- the impact of ACE, DFE, and the hub-and-spoke environment.

---

[39] This interviewee suggested that there might be a political calculation involved with contractors, given that they make it appear that the military has fewer boots on the ground.

Agile Posture and Multi-Capable Airmen

Challenges related to posture affect the AO community's ability to access manpower. Fifteen out of 19 interviewees raised this issue, and 16 insights were gleaned from strategy-vision bridging sessions related to AO community posture. One interviewee stated that "being overly focused on in-garrison operations means that MOBs are well manned, but the AO community struggles to posture their manpower in an agile way that is conducive to small teams running airfields." Furthermore, participants in the strategy-vision bridging sessions mentioned that

> We are trying to present forces for global force management in a manner similar to the other branches, but we are not organized, and we do not employ, like the other branches.
>
> Airfield Operations is not postured for a peer fight. A complete force presentation reset is required.

Another issue is that AO personnel are often not deployed as teams. Multiple interviewees and strategy-vision participants mentioned a preference for a teams-based deployment strategy, and that innovative strategies could be used to overcome current obstacles to deploying as a wing. For example,

> We crowdsource and don't deploy as a team. When I deploy, I arrive with no one else I know. I have to get the new team to work together and deal with personality issues, but it's really hard when everyone comes in separately and people come in and out every month, so we are in constant state of readjusting.

This type of deployment provides multiple benefits, but, according to one interviewee, the Air Force may not be able to implement it:

> Benefits include familiarity and team cohesion when deployed. However, [AO is] not organized, trained, or equipped to support a team-based deployment strategy.

Pervasive Equipment-Related Challenges Are Compounded by Other Factors

An overwhelming majority of our interviewees reported that AO equipment is too old, too heavy, or obsolete. There is not enough equipment in some critical areas, particularly for ACE. For ACE, equipment will likely need to be more agile and portable, but this presents some challenges for active-duty personnel. Overall, equipment-related concerns were mentioned in 18 out of 19 interviews, making it the most frequently mentioned capability gap. Nine equipment concerns were also mentioned during strategy-vision bridging sessions. One interviewee mentioned that "DFE and ACE will really affect the AO community in a negative way because that's not what we are used to."

Interviewees cited concerns regarding ownership and deployability of equipment:

> Most of the deployable equipment is in the Guard, ~20 percent [is] in the 53rd—our only active-duty deployable unit; some is left over from the Cold War, a little bit in COCOM/USAFE. CRGs have very little, TACANs etc. . . . and that's why we're facing it right now—we cannot do DFE with the way we're organized.

> We're wasting time with respect to equipment—we've known since 2004 that our deployable [navigational aids] are out of date.

AO Is Not Organized to Support ACE and DFE Demand in the Hub-and-Spoke Environment

Interviewees and strategy-vision session participants mentioned that manpower and equipment gaps affect the AO community's ability to adapt to ACE, DFE, and the hub-and-spoke environment. The topic of the impact of ACE and DFE elicited a range of responses, including five insights from strategy-vision bridging sessions and 16 interviewees.

> We are moving forward on ACE and DFE outside of the U.S., but [we have] no clear understanding of what our in-garrison requirements in contingency state look like.
>
> ACE, DFE, and the hub-and-spoke will require a lot of international agreements and deconfliction with host militaries.
>
> The problem is they don't have the ability to fight DFE or [operations plan] execution because they don't have the numbers or training.
>
> We cannot do DFE with the way we're organized.
>
> A reorganization needs to happen, because what we are doing now does not support the new NDS imperatives of DFE and ACE.

To adapt to ACE and DFE, the AO community is developing force packages and assessing ways to determine the true demand. During the strategy-vision bridging sessions, these topics have been raised consistently. As one interviewee noted, "just because their war plan says they need 15 XY and 10 YZ doesn't mean we can buy everything for everybody. [AO is] not designed to supply two war efforts simultaneously."

Lack of equipment is especially a problem for DFE and ACE. As discussed in the earlier equipment section, there is not enough of it, and what does exist is not light, lean, or mobile. For example, deployable navigational aids—which are necessary in austere locations—are continually underfunded, and equipment at host nation airfields could be targeted, which necessitates resilient capabilities. Furthermore, ACE and DFE increase the need for portable equipment, which the AO community does not possess. Rather, much of the deployable radar capability resides within the ANG, and the deployable equipment that does exist within the AO community is often too old, in that, as one interviewee said, it is not "light, lean, and mobile" even if it is functional.

Force Generation Capability Challenges

Three issue areas emerged within the force generation category of concern:

- accessibility and training
- surge capacity and backfilling positions
- requirements.

Accessibility and Training Impacts Provision of AO Manpower

First, the AO community is sometimes unable to access enough trained airmen. Fifteen interviewees cited concerns that related to the ability to access trained personnel, which were validated by 14 insights from the strategy-vision bridging sessions. As mentioned previously in this chapter, with the current setup, overall manning sometimes looks good on paper, but airmen are not always accessible for deployment because they have not completed 7-level training that is generally required for deployment.

Furthermore, whenever personnel are pulled for deployment, it means that they are unavailable to train at home. Moreover, the large training workload in CONUS facilities takes so much time that many 3-level ATC personnel at CONUS garrison facilities are training to increase their skill level. While manpower may appear adequate in the aggregate, many personnel are not accessible for deployment. Reflecting the view of many interviewees and strategy-vision bridging session participants, one SME explained that

> The AO career field went to single person UTCs [unit type codes] so they don't destroy the home station when they deploy. A unit can give one 7-level but not a whole team of 7-levels for a deployment and still be able to conduct their home mission.

SMEs commented that the number of personnel who can be deployed is limited, and it would often be preferable to keep them at home to train and mentor junior enlisted personnel. The result is that readiness may show "green" even when there are personnel limitations.[40] Level 3 ATCs have often spent so much time focusing on upgrading their training that they forget some of their foundational basics.[41]

Multiple interviewees mentioned that ACE and DFE will affect the training pipeline, whether personnel are training in-garrison to be accessible for deployment or are sent somewhere to train. Beyond the lack of enough people to cover both training and manning shortages, it takes time to move people back and forth between positions. Interviewees stated that while "there are a lot of in-garrison support exercises," it is difficult to organize, train, and equip without "breaking the base."

Many Guard interviewees believe that they are postured well to support the four-bin force generation model, but that active-duty airfield operators are not. However, interviewees within the active-duty AO community reported drastically different perceptions. For instance, active-duty interviewees felt that it is time for ANG ATC squadrons to develop the capability to provide AO services as well.

---

[40] While not mentioned by interviewees, this limitation might be addressed in the *capability readiness* reporting in DRRS-S, which is a commander's subjective assessment of the unit's ability to accomplish its designed mission (AFI 11-201, 2018, p. 5).

[41] "It takes nine to 18 months after technical training to be good in [ATC], and two years to 'certify.' . . . It takes 2.5 years to advance from a 3-level to a 5-level."

Need for Scalability to Address Surge Capacity and Backfilling Demand

Fourteen of 19 interviewees cited concerns about "backfilling" positions that are vacated when active-duty personnel are deployed. While interviewees acknowledged the need for backfilling positions, they also mentioned concerns, particularly when looking at the ANG or reserve for surge capacity:

> In the future we will have to look at [using Guard or reserve for backfilling], but a lot of our Guards are tied to homeland security missions or the FAA. . . .

In addition, interviewees mentioned "scalability" as it relates to ensuring that enduring missions do not break the in-garrison mission:

> There needs to be a process at the [career field management] level that increases awareness and prepares to act in cases where something can be done to meet requirements, especially for deployment. While 13M has been able to meet almost 90 percent of the demand, increasing scalability is still a concern.

As for equipment, an interviewee mentioned that major commands (MAJCOMs) believe that they are underequipped in some critical mission areas in relation to demand signals from war plans.

Misalignment of In-Garrison and Deployed Mission Requirements

Requirements-related concerns were mentioned in 16 of 19 interviews, and the strategy-vision bridging sessions captured 11 corresponding insights. Part of USCENTCOM's mission is to provide the same AO services abroad as at home; if U.S. Air Forces Central requirements for military personnel remain similar to what they are today, it may be difficult to support ACE and enduring USCENTCOM requirements at the same time.[42]

The constant homeland security mission in certain locations makes it even more difficult for the AO community to meet requirements. As these missions adjust, and some units go OCONUS, other units will have to pick up the slack. In peacetime and contingency states, home station requirements must be met during deployment as well. In addition, deploying to places like USCENTCOM and U.S. Southern Command is essentially like running another home station somewhere else. As one interviewee said,

> the Army does pick up and deploy all at once where we can't do that, and we have to spread ourselves out . . . got to scratch and claw every single cycle to get manpower needs filled . . . no one does the analysis to get the big picture . . . one guy in AFPC tries to get all of this together . . . Army has an easier time with this because they deploy all together and their training is all together.

---

[42] The withdrawal from Afghanistan occurred after this project was completed. The impact of the withdrawal on USCENTCOM demands for AO personnel is not addressed in this report.

Culture, Community Alignment, and the Pace of Policy Change

An interviewee mentioned that operating out of host nations requires finding out how the host nation rules differ from FAA, AO, and international standards. In addition, "13Ms need to be prepared to adapt to a more agile environment and balance friction between efficient and effective operations and safety." There are perceptions that the FAA way of doing things has become too ingrained in the AO community, and that this ought to be set aside when dealing with contingencies. One OSS commander commented that the AO community has a "garrison game"—it remains structured and equipped to perform organize, train, and equip functions that support operations at a home base: Air traffic controllers are guided by FAA standards that may not apply at deployed locations, airfield managers focus on the needs of established airfields, and SMEs indicated that RAWS personnel can work only on equipment that they are trained to use day-to-day at their home station. As one interviewee said,

> [We] will have to build and develop 13M to be able to negotiate with those host nations and be SMEs in joint force operations on what we are going to be bringing into that air space and negotiate with them that they are willing to accommodate.

Another theme within this category of concern is the no-risk mindset. Interviewees note that this negatively affects the ability to train multi-capable airmen. This is an issue of culture, particularly as it relates to the ACE construct, DFE support, and the AO community's relationship with FAA.

> One of the problems that we have is the pace of policy change. . . . Our FAA doc is from 1969 and we are slow to come to agreement with DoD and FAA on how to update that.

During the strategy-vision bridging sessions, another category of concern emerged, which was also supported by insights gleaned during our interviews. Specifically, concerns related to AO community culture and a perceived misalignment of AO within the broader USAF enterprise (that is, whether AO should function under operations or logistics) were tied to gaps in the Air Force's ability to balance the pace of policy change with the pace of emerging missions.[43] For example, interviewees and strategy-vision participants stated that

> 1C8s should be under A4 (logistics) instead of A3 (operations) because they're mostly maintenance.
>
> Practice and execution are aligned with A3, but AO is doctrinally aligned with A4.
>
> ACE development through A5 is clashing with the immediate concern of A3OD, which is developing a force presentation that addresses current needs.

However, not all interviewees agree that AO is better aligned under A4. For example, one interviewee claimed,

---

[43] This perception was discussed in the strategy-vision bridging sessions.

> We are properly aligned under the A3 and underneath the operations group at the installations level. We are a functional area of the A3 even though many still refer to air ops as base or mission support entity.

We explore this challenge of misalignment further and, in particular, how it affects developing doctrine for AO in the next section.

Reimagining Airfield Operations Doctrine

The perceived misalignment of AO within the USAF enterprise was an important topic in the Vision and Bridging Strategy discussions and highlighted the need for the AO community to be able to better see itself as a community, and for external stakeholders to better see inside of the community to make resource allocation and employment decisions. The essence of this concern centers on two questions:

1. Where does the community functionally fit from a doctrinal perspective?
2. What are the employment demands on this functionally unique community?

We looked to the Joint Publication series to better understand where the community fits doctrinally and used joint concepts to develop a characterization framework to enable the community to better understand and articulate its demands.

The AO community exists across multiple functional domains, and its personnel are therefore involved in any operation that requires an airfield or the need to land aircraft. The functional domains represented in the community primarily include operations, command and control, and logistics, with emphasis on logistics in accordance with the functional delineations set forth in the Joint Publication series. Given this diversity of domains, AO equities need to be discussed at the earliest stages of planning, which nests the community deeply into the command and control and operational lanes as well.[44]

DFE is meant to allow proactive and scalable employment of forces, which implies the ability to disperse forces. DFE is also intended to enhance the ability to compete and provide freedom of maneuver. These concepts are achieved through concentration and deal with how forces are positioned.[45]

How and where to position forces is a multifaceted problem that requires expertise in both operations and logistics. This fact illuminates the need for the AO community to develop a vision and set of capabilities to provide to the joint fight that address the operational and logistical

---

[44] Joint publications that relate to AO include the 3-0 series (operations), the 4-0 series (logistics/support), and the 6-0 series (communications). Air Force doctrinal publications are Air Force doctrine Annex 3-52 (*Airspace Control*), Air Force doctrine Annex 4-0 (*Combat Support*), and AFTTP 3-4.4 (*Contingency Airfield Operations*).

[45] DFE has some parallels to maritime theory as discussed in Julian Corbett's *Some Principles of Maritime Strategy*. Maintaining the capability for major combat relates to Corbett's idea of a *fleet in being*, which is a defensive concept that actively denies offensive gains by an enemy and ensures a force is capable of decisive action when conditions are favorable. Maintaining a fleet in being requires the force to disperse when offensive action is not appropriate and concentrate when decisive action is required (Corbett, 1911, p. 103).

aspects of the strategic end. Understanding that DFE may require dispersal of forces to achieve its stated objectives is useful to both the campaign planner and the service seeking to make decisions with respect to organizing, training, and equipping for mission execution. These are the connections that build the understanding necessary to formulate assumptions and assume risks as part of mission planning.

Linking Doctrine to Airfield Operations in the Contemporary Operating Environment

AO is a function that serves both the combat and support elements of the Air Force. As we have seen, the Air Force's strategic adjustments to doctrine and planning appear to increase the demand for AO personnel and duties.[46] These demands are largely created by the logistical realities of future operations. The AO community must understand and be optimally integrated into both the operations to be conducted and the logistics to make those operations happen.

Understanding what operations will be conducted both in garrison and while deployed is imperative to effective resource allocation and mission execution. The AO community will also need to understand the basing footprint for an operation to effectively and efficiently manage enterprise-wide AO—a function that is global and persistent throughout all types of operations. An understanding of these elements—the strategic environment, the operational requirements, and the basing footprint for an operation—creates the context necessary to construct alternatives to support AO.

Most Air Force capabilities require some form of airfield from which to operate. Correspondingly, the planning for these airfields should be a key aspect of command and control and operational strategy and planning. These airfields could be either on permanently established air bases in CONUS, in enduring locations overseas, or at temporarily established overseas contingency locations (U.S. Joint Chiefs of Staff, 2019, p. vii). The AO community needs to be aware of changes in the demand for contingency basing.[47]

According to Joint Publication 4-04, geographic combatant commanders develop a basing strategy as part of their strategic estimates, strategies, and plans to accomplish their mission (U.S. Joint Chiefs of Staff, 2019, p. vi-1). The basing strategy provides an overall arrangement of bases and how contingency locations will be used (U.S. Joint Chiefs of Staff, 2019, p. iii-7). Multiple key personnel at a contingency location are responsible for different support aspects to the base. The Senior Airfield Authority is the key person responsible for the services provided by AO personnel (U.S. Joint Chiefs of Staff, 2019, p. iii-7).[48] Understanding what service is likely to fulfill the Senior Airfield Authority role, as well as what capabilities may be provided by joint,

---

[46] Informed by interviews with AO community members and SMEs. Also see Mills et al., 2020.

[47] Joint Publication 4-04, 2019, and Air Force Doctrine Publication 4-0, 2020, are the applicable doctrine documents.

[48] The Senior Airfield Authority is appointed by the combatant commander or the Joint Force Commander.

coalition, or host nation personnel is important for the AO community to estimate the demand that it could face in each operation.

Figure 4.1 depicts one way in which the AO community could see itself in terms of roles and employment. The figure shows the functional overlap between operations and support and how AO roles are situated in that space. AO roles fall into three primary areas: (1) staff roles as special staff members or liaisons needed for planning, coordination, oversight, and reporting of AO capabilities across the planning to execution cycles;[49] (2) command and control; and (3) AO functional skills to be performed at the point of execution at an airfield, such as air traffic management.

Figure 4.1. Employment of Airfield Operations Capabilities Across Warfighting Functions

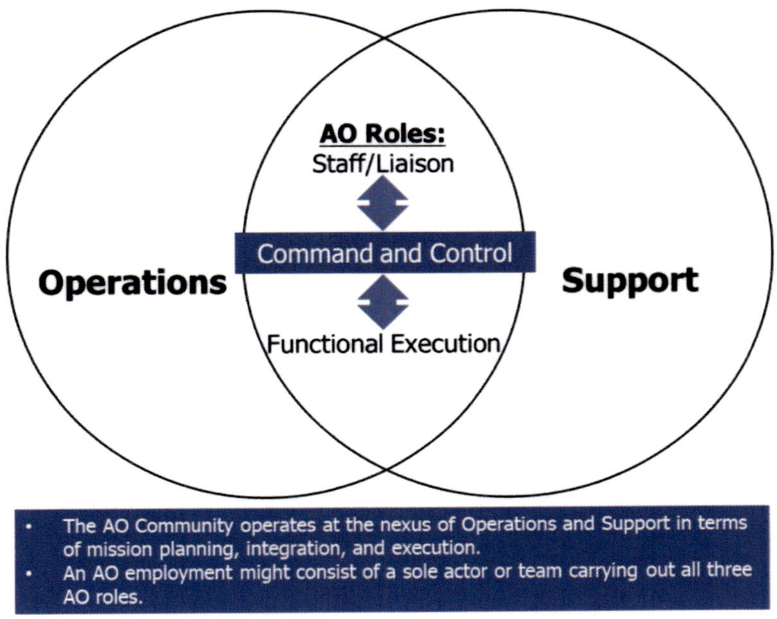

Command and control splits across two lines of effort. The first is integrating into higher headquarters staffs for liaison and planning roles, and the second is command and control of AO capabilities performing their functional requirements. The AO community must have functional operators out doing career field tasks, which require command and control during execution. Correspondingly, higher headquarters must also have staff coordination, planning, and oversight. Each of these command and control activities is, in effect, a separate capability that will be needed to meet a joint warfighter demand—even though the same individual might perform more than one of these functions when and where appropriate.

---

[49] Liaison roles can be staff or command and control in nature, and staffs exist in headquarters to support command and control tasks but are still worthy of separate lines of effort in terms of functional tracking.

Support versus operations (or combat) may be perceived by some as a trivial debate, but the distinction can have major implications. Militaries throughout the world associate more military importance with fighting-related operations and, at times, subordinate the support functions. If a community is best aligned with the macro support community for resource allocation and upkeep of community equities, shunning that community for lack of "gravitas" is shortsighted and could be costly if it results in the community not being optimally integrated or resourced.

Rather than choosing a side, the AO community should embrace its existence in the base operating system community, which is inherently support, while also maintaining appropriate presence in the command and control and operations functional communities because AO truly exists in all three. The AO community should determine how to leverage the resource allocation and higher headquarters alignment benefits garnered from the support community while keeping its presence at the right places and times for operational impact on the operations side.

5. Tools That Can Enhance Management of AO

In Chapter 2 we discussed the capabilities of the AO career field and the supply of, and demand for, AO personnel. We highlighted some of the unusual demands placed on the AO community because of in-garrison responsibilities that continue when the nation is in a contingency state, as well as other demands (such as maintaining air traffic control at civilian airfields) that are not widely known. Chapter 4 described existing challenges with AO capabilities and anticipated difficulties that the community could face as the Air Force transitions to ACE-like operations and DFE.

Among the concerns raised in the previous chapters is considerable uncertainty about the AO community's ability to satisfy the types of demands that might arise from implementation of the ACE concept. Moreover, there doesn't appear to be a ready means of evaluating where shortfalls might arise, how significant they are, and possible alternatives for mitigating shortfalls. To illustrate the value of assessment capabilities that quantify potential personnel challenges, RAND PAF developed several tools that can provide insights into potential policy solutions for closing those gaps—a personnel supply and demand model and a geographic fill-rate mapping tool.

A Framework for Tracking Supply and Demand

Demands on the AO community emerge from mission requirements emanating from a complex array of stakeholders; the supply is an equally complex group of AO capabilities across the active and reserve components, as well as government civilians and contractors. These complexities make it difficult for the AO community to discuss its supply and demand equities in a cogent and holistic fashion. For example, in interviews with SMEs, we were often unable to obtain basic information, such as usage data on the number of requests made and fulfilled or requests with reclamas.

Figure 5.1 contains a strategic process map that could be used to organize the key supply and demand nodes in accordance with the flow of activities from the generation of operational demands through the life cycle tracking and fulfilment of those demands. It is important to note that operational and intelligence indicators drive the demands irrespective of capabilities of the support community to meet those demands.

Figure 5.1. The Airfield Operations Mission Demand-to-Sourcing Strategic Continuum

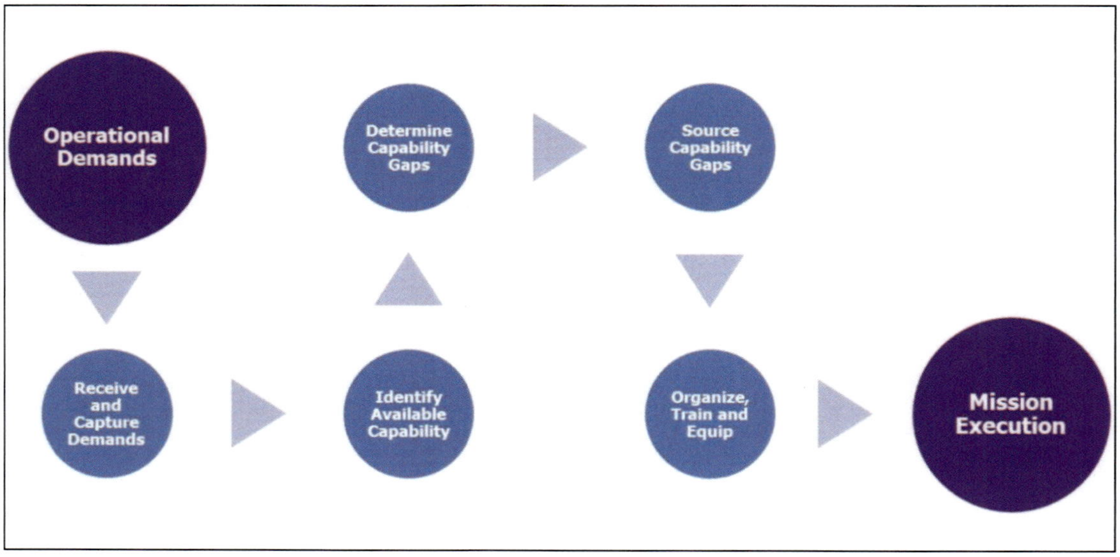

NOTE: The large purple circles are meant to represent major functional areas that are inherently joint. The blue circles represent supporting tasks accomplished by the USAF.

Figure 5.1 illustrates how the AO community could track seven key nodes to gain a better understanding of total demands for AO capabilities while subsequently working along a path to fulfilling mission requirements and maintaining visibility over capabilities in the near and future terms. Each node would include metrics and data tracking that will allow the community to have archived records of consumption data. These data are a key aspect of budgeting according to actual realized requirements and can be used to justify future adjustments to posture and force presentation of the community. Using this process map in discussions with the participants in the Vision and Bridging Strategy discussions highlighted several points:

- The inability to clearly codify or capture the expanse of the demand history is a root cause of the AO community's inability to posture or align itself for current and future operational demands.
- The mission demands that can be seen right now are being completed, but at risk to efficiency and the future health of the community.
- Some of the overall mission is not accounted for, with the MAJCOMs currently requesting what they know is available as opposed to basing requests on operational needs—and subsequently filling some of those needs through other, at times less optimal, sourcing solutions.

Within this framework, it is possible to use relatively simple tools to examine the ability of the AO community to respond to operational demands.[50]

[50] See Appendix B for a more detailed discussion of this framework and a subjective assessment of the AO community's ability to address each node.

A Supply and Demand Decision-Support Tool

RAND developed a rules-based decision support tool to help assess the ability of the AO community to satisfy the demand for personnel based on the hub-and-spoke concept for ACE-like operations that we described in Chapter 3. To start, we defined a set of demands that would be representative of ACE operational demands and then defined rules related to the ability of different sources of AO personnel to respond to those demands. This proof-of-concept model that we developed required three things:

- Data on the supply of AO personnel, including information about an individual's assigned MAJCOM, base, and location, in addition to individual attributes. Our initial analysis was limited to active-duty military personnel.
- Data on the demand for AO personnel, including the number and type of AO personnel needed at a hub or a spoke and the number of spokes supported by each hub.[51]
- A tool to connect supply and demand that allows a user to assess the ability of supply sources to support a given demand. RAND PAF developed an R Shiny–based tool that enables a user to
  - establish the demand for personnel according to the number of hubs and spokes required, as well as other demands. We also include the ability to add an air bridge support capability to support the time-phased force deployment data flow (this is a switch that can be turned on or off) and to add enduring requirements for COCOM support.
  - select sources of supply in two ways. First, a user can select a MAJCOM that is available to provide AO personnel and then select individual bases in that MAJCOM. Second, the user can identify the percentage of assigned personnel at a location that is considered available for deployment to a hub or spoke. This is a way for the user to account for *critical* levels of personnel that must remain at locations to maintain needed services.[52]
  - observe personnel surpluses and deficits in real time. The tool outputs three tables (Demand, Supply, and Personnel Health) that respectively summarize the demand (given the input of hub, spoke, and airbridge requirements); the supply of each base (given the input of the percentage available for deployment by each base); and, lastly, personnel health (the difference between the demand and supply for the 13M, 1C1, 1C7, and 1C8 career fields). For comprehensive understanding, a bar chart that visually represents the personnel health data table is output as well.[53]

---

[51] This was based on descriptions in PACAF, 2019, and discussions with SMEs about the PACAF CONEMP dated May 10, 2021.

[52] This proof-of-concept tool allows the user to specify only an overall percentage of individuals available for deployment. It would be easy to modify the tool to allow specifications by skill level to account for the possibility that, for example, only fully qualified 7-level personnel are allowed to deploy.

[53] Appendix C contains the R code for the tool.

A PACAF Example

To illustrate how the model works, we developed a scenario involving active-duty forces across the globe (and assume homeland defense demands will be met by ANG forces). We assume PACAF plans to establish a CONOP that requires one hub and four spokes. Each hub requires one 13M, five 1C1XXs, two 1C7XXs, and two 1C8XXs. A spoke requires three 1C1XXs, one 1C7XX, and one 1C8XX, but no 13Ms.[54] PACAF wants to have as little impact as possible on its bases in the Pacific, so the only sources of AO personnel for the hub and spokes are its two bases in Alaska. Critical manning at Eielson AFB and Joint Base Elmendorf-Richardson is 75 percent, so each base can relinquish only 25 percent of its AO personnel for the hub and spokes.[55]

Figure 5.2 shows how the setup for this example appears in the model. The user establishes the demand by specifying one hub and four spokes, then selects PACAF as the MAJCOM source of supply with only Eielson AFB and Joint Base Elmendorf-Richardson selected with the "sliders" for each base set to 25 percent. In the figure, the bar chart and the table labeled "Personnel Health" depict the supply and demand per generalized career field. The supply of 13Ms is insufficient (as shown by the red bar for 13M at the left of the bar chart) with a deficit of one. There is also a deficit of seven for 1C1 personnel (as indicated by the second red bar). The green bars for 1C7s and 1C8s show that these career fields can satisfy the hub-and-spoke demand and still have personnel left over.

---

[54] Our data indicate five of the seven CRSs (the 321st, 435th, 621st, 821st and 921st) are composed similarly to the proposed PACAF hubs (e.g., the 321st is currently assigned one 13M, four 1C1s, two 1C7s, and two 1C8s). However, there is no indication that CRSs will be used to provide hubs or spokes.

[55] The model uses a "floor" in its calculations. That is, if a location has four 1C1 personnel, 25-percent availability would mean that it could provide 1 person for a hub or spoke. However, if a location has three personnel, 25 percent would be 0.75 people, which the model counts as 0 available.

Figure 5.2. PACAF Example: Initial Calculation

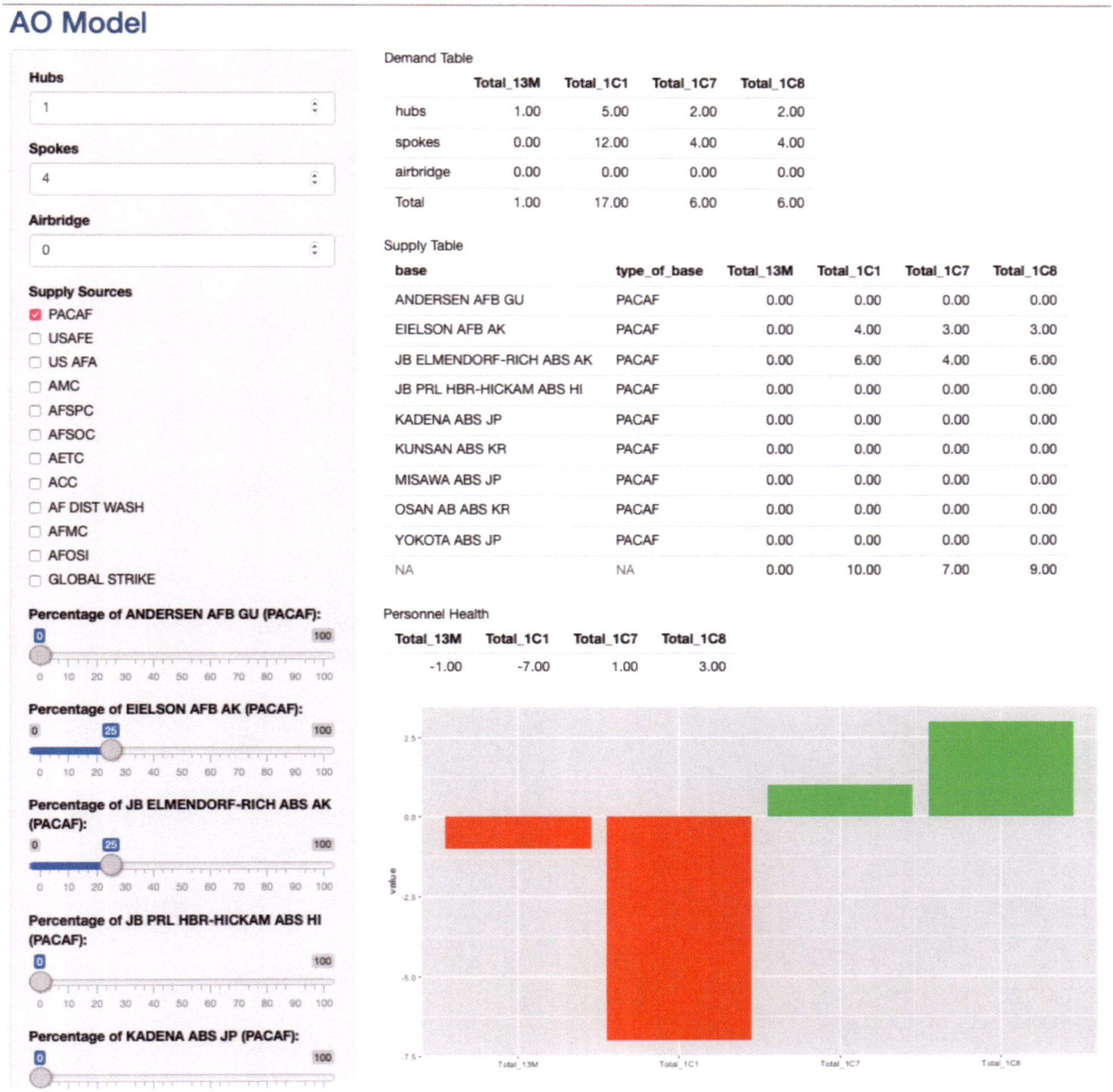

SOURCE: Screenshot of RAND PAF AO Model.

Recognizing the potential shortages of 13Ms and 1C1s, the model allows the user to determine other sources of personnel. Figure 5.3 shows what happens if Joint Base Pearl Harbor–Hickam and Yokota Air Base are also possible sources of personnel in addition to the Alaska bases (allowing use of 25 percent of their AO personnel), while still fencing off Korea as a supply source.

Figure 5.3. PACAF Example: Second Calculation

SOURCE: Screenshot of RAND PAF AO model.

With these changes, the green bars in Figure 5.3 make it appear that the supply is more than sufficient to meet the demand for the enlisted career fields but that, interestingly, the need for

13Ms is just met. In this case, Joint Base Pearl Harbor–Hickam has four 13Ms available and can therefore provide one person to make up the deficit, while Yokota Air Base has only three 13M personal assigned, and it cannot provide anyone (25 percent of three is 0.75). But there might be other ways to satisfy the demand while still restricting PACAF supply sources to the Alaska bases by, for example, allowing a larger percentage of personnel to be taken from the Alaska bases or reaching out to non-PACAF locations. The model allows the examination of other options and, in the course of exploring them, should remind the user that there may be missions at Air Combat Command (ACC), Air Force Materiel Command, or AMC bases that may not be executed if people are taken to fill contingency demands.

A PACAF Plus USAFE Example

The NDS delineates the need to be ready for the possibility of facing demands in more than one theater of operations simultaneously, so we use the model to show the potential impact of requirements in PACAF and USAFE at the same time, with two hubs and eight spokes in each theater (four hubs and 16 spokes total). While USAFE has not yet clearly defined the requirements for its implementation of the hub-and-spoke concept, for this example we assume that the hub-and-spoke demands are the same as for PACAF. We make the following additional initial assumptions:

- PACAF "fences off" all bases except Alaska, allowing 25-percent deployment of those personnel
- USAFE "fences off" Ramstein and Incirlik, allowing 25-percent deployment from other USAFE locations
- One air bridge is used for logistics support.[56]

The results in Figure 5.4 show that, for the requirements in this example, there is a deficit of one for 13Ms but surpluses for 1C1, 1C7, and 1C8 personnel. A user can easily adjust settings for other locations.

[56] SMEs at AMC suggested an air bridge would require four 1C1, four 1C7, and two 1C8 personnel at intermediate airfields.

Figure 5.4. PACAF Plus USAFE

SOURCE: Screenshot of RAND PAF AO model.
NOTE: "Sliders" are not shown for bases alphabetically after Mildenhall.

While this model can reveal potential gaps, it has several limitations. First, it deals with only active-duty personnel—though it can encourage consideration of where the reserve component, civilian personnel, and contractors might have a role. Second, it may not properly partition people by their assigned commands—that is, it may allow USAFE personnel to fill gaps in PACAF, which is probably not what would happen in practice.[57] Finally, this tool assumes that everyone is available for deployment, but the AFFORGEN construct may only make one-quarter of the people assigned to a base available, or only some subset of bases available, because of the desired binning structure.

Nonetheless, shortages such as those indicated in these examples prompt several questions:

- What additional active-duty personnel sources could be used to fill the gaps in supply?
  - Are critical manning levels at all locations understood well enough to allow planners to use them to fill hub-and-spoke shortages?
- Looking to the future, are there alternative sources of supply?
  - Can reserve forces be used for some of the hub-and-spoke demand, or could they be used for home-station requirements to free up more active-duty personnel for use at hubs and spokes?
  - Both examples indicate that 13M, 1C1, and 1C7 career fields encounter shortages initially, but that the supply of 1C8 personnel is fine. Does this suggest opportunities for MCA training that might allow 1C8 personnel to fill other AO hub-and-spoke requirements? Or is the demand for 1C8 correct when looking at hub-and-spoke operations?
  - Is the current plan for the composition of hubs and spokes the "right" one given the skills required and the plans for future AO personnel requirements?

Our prototype model with this type of fidelity allowed us to examine these and other questions and can help inform decisionmakers; a more formal, officially sanctioned model would appear to benefit the community.

A Location and Fill-Rate Map

We quickly mention another tool that can improve the understanding of the nature of the AO supply. Using the same data source as the one used for the supply and demand model and additional information about *authorized* positions, we created maps that show the location of AO personnel and the fill rates of those locations. Figure 5.5 is one example for CONUS locations. The circles on the map indicate Air Force locations with AO personnel, and the sizes of the circles increase as the size of the active-duty, permanent party population increases. In this example, circles are color-coded to indicate the percentage of authorized personnel who are assigned to the location. For example, Grand Forks has 51 assigned but 75 authorized, for a fill

[57] As of August 2021, the Air Force is still developing approaches to partitioning forces to support AFFORGEN.

rate of 68 percent, which colors the circle light orange. At the other end of the spectrum, Mountain Home, with a larger circle indicating a larger population, is manned at 87 percent and is colored green. Shaw AFB in South Carolina is in between in size and in fill rate, at 73 percent.

Figure 5.5. CONUS Location AO Personnel Fill Rate

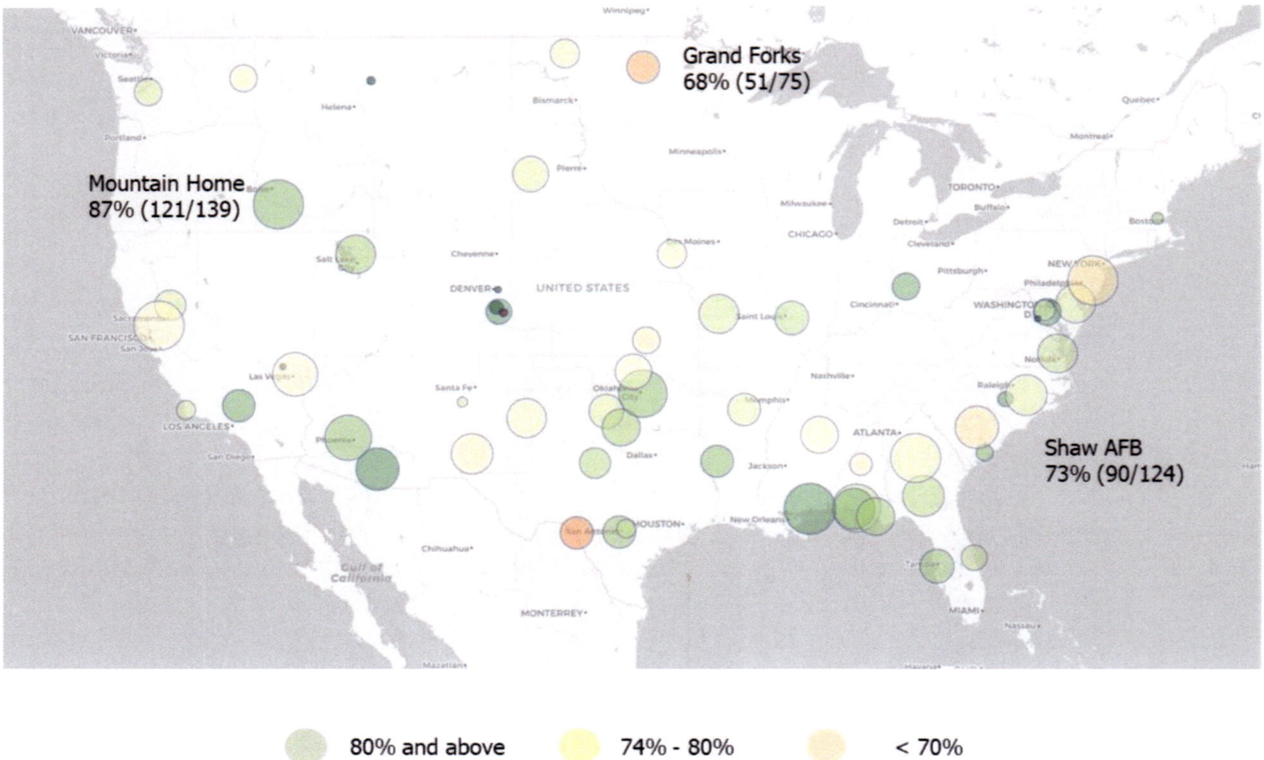

SOURCE: Screenshot of RAND PAF AO location mapping tool.
NOTE: The tool shows maps of worldwide locations and fill rates; CONUS locations are used for illustrative purposes. In the tool, if a cursor is place over the circle that represents a location, the number of assigned and authorized personnel and the percentage fill rate appear. As fill rates increase above 80 percent, circles get "greener"; as they go below 70 percent they turn more orange.

A simple example of the potential usefulness of this visualization is to place markers for ANG bases on the map at the same time, as in Figure 5.6.[58] The active-duty locations in the figure are the same as in Figure 5.5. The blue circles represent the locations of ANG AO personnel and the relative sizes of their populations. Note that the proximity of Shaw AFB to McEntire Joint National Guard Base is immediately obvious and suggests at least two things. If the 75-percent "standard" for critical manning is used, Shaw AFB appears to be undermanned. If this is affecting the AO services provided at Shaw AFB, is there a way for the ANG to supplement them? Second, would it be possible for McEntire Joint National Guard Base to assist

[58] The tool can provide the same information for civilian AO personnel.

Shaw AFB during deployments? For example, if ANG personnel could fill in for deployed active-duty personnel, Shaw might be able to provide more people for hub-and-spoke operations, and ANG personnel could provide services without being deployed.

Figure 5.6. CONUS Location AO Personnel Fill Rate, Adding ANG Locations

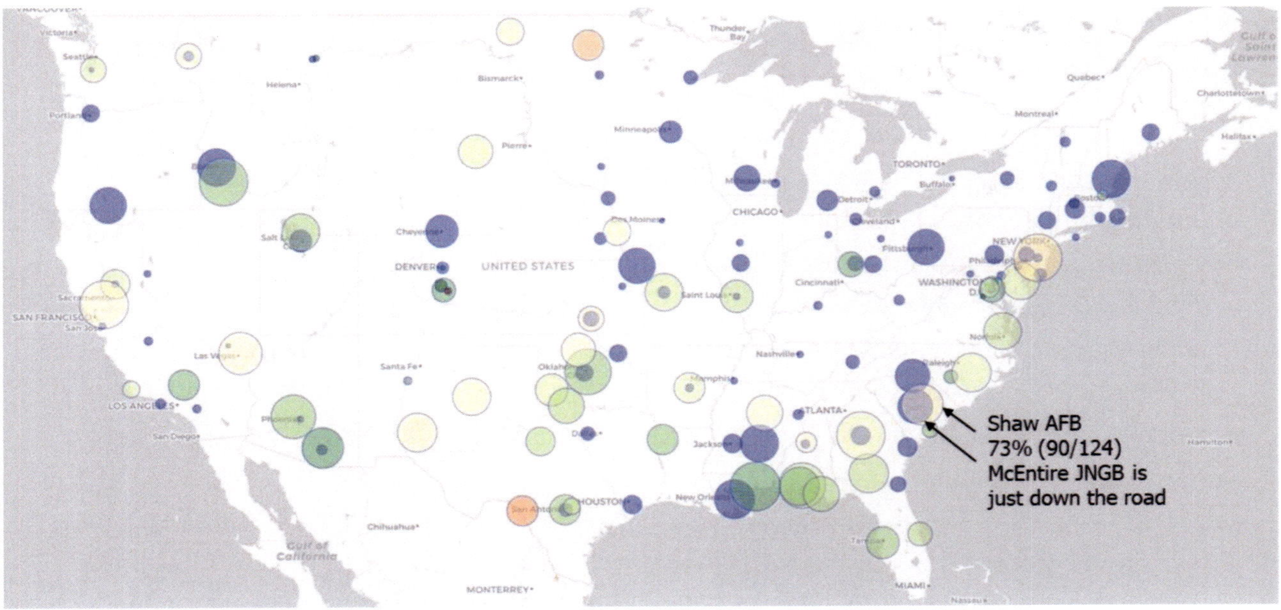

SOURCE: Screenshot of RAND PAF AO location mapping tool. Blue circles represent ANG locations, with larger circles indicating larger numbers of ANG personnel.

Conclusion

Airfield Operations-Specific Issues

Using "rule-based decision support tools,"[59] such as the RAND PAF AO model, we were able to show that, with the assumptions underlying the tool, the Air Force needs to carefully examine the demand that a hub-and-spoke concept will put on the AO community and consider the implications for sources of supply and the composition of personnel in the hubs and spokes.[60] The AO community is already developing such tools as the Airfield Operations Data Management System to track certifications of ATC personnel; an "official" version of the AO tool could also be useful for exploring organizational and deployment options in the ACE and DFE environment.

---

[59] This term was used by our late RAND colleague C. Robert Roll.

[60] We also note that this tool focuses on personnel and does not show the potential shortages in the supply of equipment that were mentioned by SMEs in our interviews.

In our discussions with SMEs, we learned that many did not have a sense of the distribution of AO personnel and specific locations that might have *overall* personnel shortages. The map visualization tool we developed provides a source of insight into this information.

Results from these prototype tools appear to confirm concerns expressed by those we interviewed about the ability of the AO career field to satisfy new demands resulting from implementation of the ACE concept.

Air Force–Wide Issues

It is important to note that many of the challenges for the AO community described in Chapter 4 have arisen in the past for other career fields, and other RAND studies have highlighted the importance of understanding personnel supply and demand issues as discussed in this chapter. For example, Dahlman, Kerchner, and Thaler, 2002, and Conley et al., 2006, analyzed data on the determination of manpower requirements and showed that the USAF tends to underestimate the time required to conduct on-the-job training (OJT). Conley and colleagues also stressed that skill-level imbalances (too many personnel at lower grades relative to the number of middle-grade trainers) can increase the burden of OJT and interfere with other mission activities. Maintaining OJT during times of conflict could limit the availability of more senior AO personnel for ACE missions—a supply constraint that could be accommodated by the simple model presented in this chapter.

When the USAF was refining its Air Expeditionary Force policies in the early 2000s, Snyder et al., 2006, developed the Deployment Capabilities Assessment Tool (DCAT), a sophisticated tool that helps identify factors that limit deployment capabilities and provides an analytic basis for balancing resources, and Mills et al., 2014, addressing new USAF missions that "entail operating in increasingly austere environments and under extremely high threats, often with dispersed forces, and sometimes outside the traditional boundaries of air bases" (Mills et al., 2014, p. 2), studied alternative approaches to determining manpower requirements to fulfill those missions.

Finally, Robbert et al., 2014, examined the impact of allowing more latitude in using reserve component forces for active missions.

As it considers the potential courses of action described in the next chapter, the AO community can also derive insights from the studies mentioned above.

6. Potential Courses of Action

In the previous chapters, we highlighted a variety of challenges that the AO community will face (and, to a certain degree, already faces) as the Air Force adjusts to DFE, develops CONOPs consistent with ACE, and uses the new AFFORGEN approach to force generation. The goal is to create a "ready, trained and equipped Airfield Operations force . . . fully integrated with our coalition partners and sister services" (Wilson, 2021). To achieve these ends, the AO community needs to take steps to improve the management of AO resources and its understanding of the supply of and demand for AO capabilities. This chapter summarizes our observations and recommendations in these two important areas.

The Supply of and Demand for Airfield Operations Capabilities

Observation: Viewing AO capabilities Through Supply and Demand Lenses Is Important

We struggled in this project to quantify the complex demands for AO capabilities. The matrix used in Chapter 2 highlights the fact that AO personnel have numerous responsibilities that must be fulfilled at their home stations even if contingency operations require deployment of AO capabilities:

- support for homeland defense
- tanker and cargo air bridge support during a contingency
- home station installation AO support to ensure the continuation of pilot training and pilot upgrade and preparation for deployment
- continued support at AMC home stations to allow aircraft to rotate through for heavy maintenance
- support for Global Strike missions that are conducted from CONUS
- ongoing support to Air Force Materiel Command to enable delivery of personnel and equipment resources to warfighters
- support for Air Education and Training Command to continue to produce aircrew.

What this means is that just because a person is coded deployable does not mean that the unit can afford to let that individual deploy at a time of need. As shown in Chapter 5, new demands arising from future CONOPs and the desire to maneuver forces during a conflict potentially create the need for additional AO capacity for both departure and arrival airfields—or a reassessment of the structure of hub-and-spoke operations. Clearly articulating home station demand is necessary to justify deployability coding of individuals and the readiness status of units with AO personnel.

Recommendation: Designate a champion to monitor overall demand.

One approach to better understanding overall demand for the AO community would be to designate a *champion* or proponent—an office or organization—to track, quantify, analyze, prioritize, and publish demands for the AO community, perhaps through the process described in Figure 5.1. Possible organizations to fill this role include the following:

- Headquarters Air Force, War Planning and Policy Division (HAF/A3O)
- ACC functional area manager
- AFFSA
- 53rd ATC Squadron.

The responsible agencies will need to be read into existing and future operational plans, CONOPs, and CONEMPs. There are precedents for developing representative data at the collateral level (for example, omitting highly classified data, such as quantities and locations, that may be important to the theater planner but are not needed for AO analysis) to provide visibility into the overall demand for the AO community.

The champion (and an associated analysis team) will need to know what other AO capabilities are available at the different OCONUS locations, as host nation or other services' capabilities planned for some locations could be tapped. Using Japan as an example, the AO personnel are most likely going to be working with existing Japanese capabilities to fulfill responsibilities, so there may be fewer or different AO requirements. The champion should have this enterprise view of the AORs to understand what types of AO capabilities are needed where.

Observation: Even If All AO Demands Were Well Understood, It Is Not Always Clear How Decisions Are Made to Align Available Supply with Demand

The Air Force has four "special" sources of airfield operations expertise:

- Airfield operations flights in OSSs, which provide ATC services, airfield management services, and RAWS equipment management on a day-to-day basis. Individuals in these organizations are deployed—though usually as individuals—to provide these services at deployed locations.
- CRGs, which are designed to open forward contingency air bases and conduct air support operations anywhere in the world. SMEs believe that CRGs currently have a mobility focus—though there are plans to change to a "full spectrum" approach for both mobility air forces and combat air forces operations (Schmidt, 2020).
- The 53rd ATC Squadron, which provides deployable ATCALS to combatant commanders when they need to establish a new airfield, project power into a contested area, or supplement existing ATC systems.
- Ten ANG ATC squadrons, which have the mission of providing rapid deployment, operation, and maintenance of ATCALS to ensure safe, efficient flying operations in both deployed and nondeployed environments.

In our discussions with SMEs, the relationships among these organizations and how their services were requested and provided was unclear. Important questions that must be addressed for future operations include the following:

- How will these four sources of AO capabilities be binned under the new AFFORGEN model, and what USAF business rules will inform sourcing decisions to meet a specific contingency AO demand signal (DFE, ACE, or other)?
- What is the relationship between the CRGs and the 53rd ATC Squadron? Who can be tasked, and by whom?[61]
- How long will it take to task the ten ANG units that own the deployable equipment? Can the USAF task the equipment and not the people? Under what circumstances? Who can do it?

It appears to us that most of these questions tend to be addressed at the time of execution and are not planned in advance with existing processes and procedures—as indicated by the number of waivers and reclamas needed to fill enduring requirements today.[62] Since the USAF has made the decision to have most of its deployable ATC capability in the ANG, it needs to establish a process—or take advantage of existing processes, if it can—to ensure that the capability is available when needed for hub-and-spoke operations. In light of the increased emphasis on agility, the USAF also needs to clarify the role for future deployments of CRGs that were created for the purpose of responding to contingency requirements. We were struck by ACE concepts that assume that ATC, airfield management, and RAWS personnel work together, when in fact those three capabilities are currently spread among distinct organizations: OSSs, CRGs, ATC squadrons, and RMCs.

Recommendation: Develop analytic tools and capabilities to provide visibility into the full demands for AO capabilities.

An important step in aligning these supply sources to future demands will be to develop and use analytic tools, such as those described in Chapter 5, to identify the available pool of forces to meet demands. Since the largest AO career field—1C1XX—has already developed the Airfield Operations Data Management System, we suggest that the 1C1 career field manager explore incorporating the AO tool from Chapter 5 into that model. Until the USAF understands the overall demands—as monitored by a designated champion—it will struggle to align available supply to fulfill the demands. It will also be important for the Air Force to maintain a "holistic"

---

[61] A reviewer of this report made the important point that the USAF may not be able to develop these rules independently: CRGs are assigned to different COCOMs that determine where and when they will be employed in coordination with the Joint Staff. AMC's CRGs are under COCOM authority of U.S. Transportation Command; PACAF's 36 CRG is under COCOM authority of U.S. Indo-Pacific Command; USAFE's 435 CRG is under COCOM authority of U.S. European Command, and elements are often shared with U.S. Africa Command. (See AFI 10-202, 2020.)

[62] AFPC provided us with information on reclamas from fiscal years 2016–2021; this information is not available to the general public.

view of the tactical, operation, and strategic vision for the AO community that will arise from future Vision and Bridging Strategy discussions.

Recommendation: Identify a decision authority with responsibility for reconciling demand for and supply of AO capabilities.

Since SMEs whom we interviewed believed that future AO demands may exceed the supply and that it is unlikely that the Air Force will fund personnel increases, a decisionmaking authority needs to be established to determine the priority for filling AO capability needs. We recommend that the Headquarters Air Force functional area manager fill this role because,[63] according to Air Force instructions,

> the functional area manager should be the most knowledgeable and experienced person within the functional area and have the widest range of visibility over functional area readiness and capability issues (AFI 36-2670, 2020, p. 313).

Developing approaches to prioritizing these demands should be part of the ongoing discussions related to AFFORGEN development, but the designated champion previously recommended would also be an important part of the prioritization process.

Observation: AO Equipment Is Outdated and May Not Be Appropriate for Hub-and-Spoke Operations

Almost all interviewees emphasized that AO equipment is outdated. In addition, while equipment demands for proposed hub-and-spoke operations are not yet fully defined (and may differ for PACAF and USAFE), the examples described in Chapter 5 indicate potential stresses on equipment availability. The PACAF "one hub and four spoke" example could require five airfield management trailers when the USAF has only seven and five mobile ATC towers when the Air Force has 17 (that are 22 years old), making it difficult or impossible to satisfy the needs of more than one hub.[64] Higher demands from USAFE and PACAF could exhaust these resources, which is another reason why clearly describing the future demand for AO capabilities is important. Once this demand is understood, the AO community will need to prioritize the supply of available resources.

Even if the warfighting MAJCOMs, Headquarters Air Force, and force providers with organize, train, and equip functions could all agree on plausible equipment solutions, acquisition timelines, and funding streams, it will take time to acquire the equipment. In the meantime, deployable capabilities in the ten ANG ATC squadrons may be required, and the USAF should

---

[63] In the experience of one of the authors of this report, functional area managers in other career fields perform this function and fill needs based on priority or reclama the requirement.

[64] This is based on potential unit type codes (UTCs) for hub-and-spoke operations that suggested "medium" and "large" AO packages would each require an airfield management trailer and a mobile ATC tower. We assume here that a hub is large and a spoke is medium. These requirements were described in an undated slide entitled *AO UTC Packages* provided by AFFSA.

determine how to best leverage those capabilities before they are actually needed. The AFFORGEN model may create opportunities for the AO community to better align this equipment to support future fights.

Recommendation: Align active-duty units with ANG ATC units and coordinate with AFFSA to procure common equipment across services.

As the Air Force develops its AFFORGEN concept, the AO community should consider aligning active-duty and other ANG units with the ten ANG ATC units that currently own deployable equipment. By doing so and ensuring that appropriate manpower is assigned to each of the four AFFORGEN phases, each phase could have 2.5 "sets" of ATC capability per phase in addition to the ATC capability that currently resides in the 53rd active-duty ATC squadron.

Aligning equipment across components would improve the utilization of current equipment but does not address the need for new deployable equipment to respond to the ACE employment approach. AFFSA is working to procure future equipment,[65] but USAFE and PACAF will need to buy deployable equipment for their hubs and spokes.

The functional area manager is "accountable for the management and oversight of all personnel *and equipment* within a specific functional area to support operational planning and execution" (AFI 36-2670, 2020, p. 313, italics added), so exploring this course of action should be the responsibility of that office. The functional area manager should also be responsible for coordinating with AFFSA to procure common equipment with other services when appropriate.[66]

Management of AO Resources

Observation: New Approaches to Managing Personnel Resources Are Needed

Our analysis of SME comments from interviews and study groups highlighted concern about whether the organization of AO capabilities will make it more difficult to satisfy the demands of the DFE strategic environment. Key personnel concerns include that the AO community has a responsibility to maintain home station operations and training even while it contributes to contingency operations. There were also concerns about implementation of the MCA concept if it is considered a way to "do more with less." As noted in Chapter 3, there is risk in aligning too many capabilities with too few people; in a high-end fight, if there is personnel attrition and an MCA is lost, the unit loses all the capabilities for which that individual was responsible.

---

[65] During discussions with SMEs, it appeared that PACAF, USAFE, and AFFSA were considering different approaches to equipment solutions and that coordinating the approaches was challenging. An AFFSA briefing to the Executive Council implied that some proposals for acquisition were not yet funded.

[66] SMEs noted that there have been deployment situations in which the Air Force has had to rely on the Army and Navy to provide equipment. As one remarked, "We're pursuing a mobile radar that the Army and Navy have already been using for 20 years."

Recommendation: Consider establishing active-associate units at high-priority CONUS locations.

The Air Force should explore approaches to using ANG personnel to replace active-duty deployed personnel at key operating locations, such as AMC bases. Aligned units could operate in a similar manner to active-associate units in that ANG units would know where they would be needed to backfill deploying active-duty personnel. Innovative approaches to using civilian personnel and contractors at CONUS locations during contingencies should also be considered to improve the availability of active and ANG personnel: Having ANG, civilian, or contractor personnel available to maintain home station requirements might free up members of the AO unit to deploy together as a capability instead of using crowdsourcing.[67] Detailed proposals for increasing the involvement of Air Force Reserve personnel have been made but, according to interviews, would be difficult to support because resources are a zero-sum game: The active-duty would need to give up personnel positions to fund reserve positions.

The establishment of active-associate units is important when the existing "special" units are taken into account. Active AO personnel might be associated with ANG ATC squadrons to train and prepare to deploy together. ANG AO personnel might be associated with active contingency response groups to improve the ATC capabilities of those groups. AO personnel assigned to OSSs could be associated with ANG ATC units or the ANG CRG to better train for contingencies and help them escape the home station "mindset" that was criticized by those we interviewed. Associate units could provide more access to ANG-owned equipment, enabling ANG to backfill active-duty to allow them to deploy earlier (assuming ANG utilization is made easier), and improve training opportunities.[68]

Observation: The New AFFORGEN Model May Provide an Opportunity to Leverage AO Capabilities to Satisfy Expeditionary Requirements

AFFORGEN divides all combat forces into the four "buckets" of reset, prepare, ready, and available to commit. In the meantime, ACC is working with the concepts of "lead wings," which deploy as the main organization in charge of a set group of aircraft from different bases and wings (Cohen, 2021).[69] Headquarters Air Force is developing the idea of "air base squadrons" to provide forces for long-term sustainment of operating locations—squadrons that, for example,

[67] There are trade-offs for these options. The use of contractors would free up manpower slots for active-duty or AFR personnel but would cost money. Use of civilians would not free up manpower slots, and it is unclear how the costs would compare. Civilians can be deployed (see Levine, 2021), but interviews indicated that the practice is not widely used, so by increasing the employment of civilians, deployable capabilities could be reduced.

[68] We acknowledge legal constraints on the use of the Reserve component—even within the new AFFORGEN model. For example, AFR and ANG units will not be postured in a "prepare to deploy" status unless activated (U.S. Air Force, 2021a, para. 6.3.2.4.5).

[69] A reviewer notes that AMC Directorate of Air, Space and Information Operations is also exploring options for AMC lead wings.

reduce support for a fighter wing from 1,000 to 200–300 people.[70] Command and control is important in this context, and, as indicated in the overlap between operations and support in Figure 4.1, ATC and airfield management should explore ways to be included in the air base squadron bucket. AFFORGEN is an opportunity to be clear about and reckon with actual AO demands and how they will be supplied.

Recommendation: Incorporate the AO community into the AFFORGEN model.

Fitting the AO community into the AFFORGEN model requires understanding the issues summarized in this chapter and making use of analytical tools, such as those described in Chapter 5, to understand the impact of the constraints of the AO supply on the ability to satisfy DFE demands. Then the supply prioritization strategies (such as incorporating active associate–like connections) and equipping strategies previously mentioned can be leveraged to posture the AO community into the AFFORGEN model.

Conclusion

The goal of the ongoing work of the *Air Force Airfield Operations Vision and Bridging Strategy* working group is to refine recommendations to "organize, train, equip and prioritize risk in support of resource allocation, future airmen training protocols, [and] employment strategies in the near and mid-term" (Wilson, 2021). As that team refines its recommendations, viewing AO capabilities through supply and demand lenses and recognizing the impact of future requirements on the management of all personnel and equipment resources will be of key importance. We believe that the recommendations in this chapter will inform the working group's deliberations.

[70] Discussion with an SME from ACC Directorate of Air, Space and Information Operations, June 8, 2021.

Appendix A. Organizations Interviewed and Approach Used for Interview Analysis

Organizations Interviewed

Interviewees were promised anonymity in our discussions, but Table A.1 lists the organizations we contacted for interviews, documents, and data.

Table A.1. Organizations Interviewed or Consulted

Organization Type	Organization
Headquarters Air Force	A3OJ: Airfield Operations Force Development
	A3OD: War Planning and Policy
	Functional managers
	Career field managers
Major command and numbered Air Force Headquarters	ACC
	AFR Command
	Air Force Materiel Command (Air Force Life Cycle Management Center)
	PACAF
	USAFE/A3CA
	USAFE – Air Forces Africa
	AFNORTH/1AF
	U.S. Air Forces Central/9AF
COCOM headquarters	USCENTCOM
	U.S. European Command
	U.S. Northern Command
	U.S. Pacific Command
Other organizations	USAF Expeditionary Center
	AFFSA FAA
	National Guard Bureau
	OSSs (75th OSS, 18th OSS) 297th ATC Squadron

Qualitative Analysis Approach for Interviews

The research team used a mixed-method phased methodological approach to analyze insights from three distinct data sources: October 2021 Executive Council briefings, structured interviews, and AO strategy-vision bridging sessions (Figure A.1).

Figure A.1. Airfield Operations Data Analysis Phased Approach

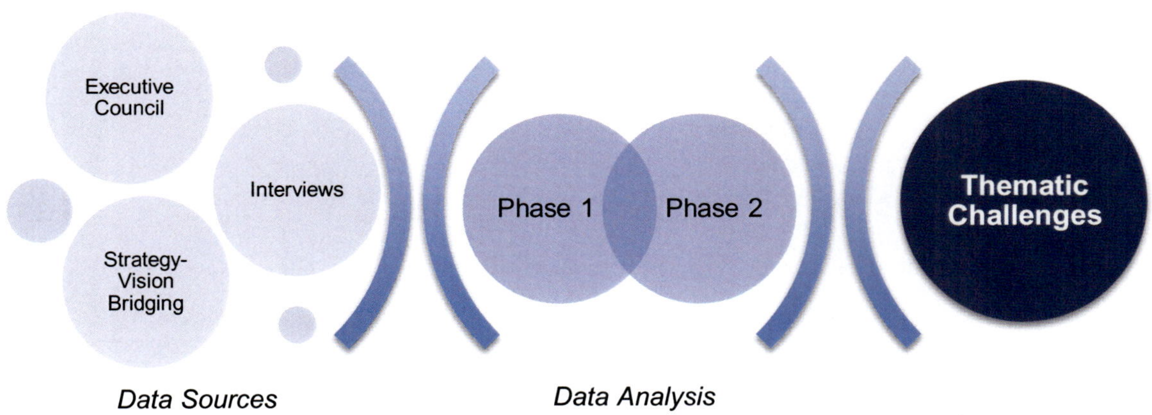

Interviews yield a very large amount of unstructured data, which makes it difficult to parse common themes and fully understand the context behind the data by simply reading notes.

During the first phase of analysis, we organized coded insights from the interviews into a modified DOTmLPF-P framework structure. Coding of interview data was done concurrently with thematic analysis and continuing interviews to develop a better understanding of which areas should be explored in more detail and what types of issues were discussed most frequently in interviews. The result allowed the research team to develop a list of thematic challenges as well as past, present, and future recommendations mentioned by participants. The same approach was used to analyze insights gleaned from the executive council briefings and the strategy-vision bridging sessions.

The second phase of analysis consisted of validation and review of the outputs from phase 1, which allowed us to identify thematic patterns across data streams and determine whether certain challenges were more prevalent in certain data sources. Table A.2 provides more detail on our phased methodological approach.

To better understand which management areas aligned with the categories of concern and thematic challenges identified in phase 1, we created an annotated framework in which we organized the themes across a modified DOTmLPF-P model to categorize secondary-level data. The elements of the annotated framework are outlined in Tables A.3 and A.4.

Table A.2. Mixed-Method Phased Data Analysis Approach

Phase	Task	Output
Phase 1: Data classification	Data coding • Issues, problems, challenges, obstacles • Contextual information • Recommendations (past and future) • Suggested literature, additional data sources, points of contact Thematic analysis	Overarching categories of concern
	Primary-level binning structure • Personnel-related themes • Force presentation themes • Force generation themes	Thematic analysis of challenges within each category of concern
	Secondary-level binning structure • Modified DOTmLPF-P++ framework analysis	Organization of challenges across management areas and frequency rates
Phase 2: Validation and review	Cross-check insights across data streams (i.e., executive council briefings, interviews, strategy-bridging sessions)	Final thematic analysis across the modified DOTmLPF-P++ framework with frequency counts

Table A.3. Primary-Level Analysis

Management Area	Manpower and Personnel	Force Presentation	Force Generation
Doctrine (D)			
Organization (O)			
Training (T)			
Materiel (m)			
Leadership and education (L)			
Personnel (P)			
Facilities (F)			
Mission (+)			
Strategic communication and coordination (+)			

Table A.4. Challenges Across Management Areas

Category of Concern	Gaps	D	O	T	M	L	P	F	P	+	+
Manpower	• Insufficient manpower; too-small career fields; crowdsourcing										
	• Impact of manpower demands on training, accession, retention, deployability, and mission execution										
	• The role of ANG, AFR, civilians, and contractors										
Force presentation	• Posture										
	• Antiquated/heavy equipment and failing infrastructures										
	• Impact of ACE and DFE; hub-and-spoke environment										
Force generation	• Accessibility and training										
	• Surge capacity and backfilling positions										
	• Requirements										
Culture/other	• Culture; community alignment; pace of policy change versus pace of emerging missions										

SOURCE: RAND Modified DOTmLPF-P Framework Analysis.

As an extra level of descriptive granularity, the research team established a list of additional relevant terms from the preliminary phases of the research with which to categorize and assess the interview data. We used keywords to better understand patterns that were emerging in the data and more easily parse sections of text by topic for qualitative analysis. Additional codes were developed as the team worked through the data in an iterative process using thematic analysis. This allowed the researchers to refine definitions and pull out as much context as possible from the data. The keywords used included the following:

- force generation; reset; prepare; ready; commit; accessibility; requirements; scalability
- force presentation; deployability; readiness; manpower; human capital; equipment; posture
- doctrine; authorities; responsibilities; mission(s); alignment; coordination; strategic communication
- demand signal(s); homeland defense, joint operations; supply; training
- joint conflict continuum; military engagement; security cooperation; deterrence; crisis response; limited contingency operations; large-scale combat operations; HADR/FHR

The framework presented in Table A.3 was used to guide the secondary-level analysis and generate frequency counts for prioritization of thematic insights.

In Figure 4.2 in the body of the report, we categorized implications of responses as *strong* or *moderate*. For the interviews, we gauged *strong implication* when more than five separate interviewees brought up the issue. For the 11 strategy-vision sessions, we gauged *strong implication* when the topic or insight was brought up in three or more separate sessions, and for the executive council briefings, we gauged *strong implication* when a topic or insight was brought up in two or more briefings (because there were fewer of those than the other two streams).

To combine the information from the three sources, we aggregated the frequency counts for all streams, which are the numbers within the DOTmLPF-P++ matrix in Figure 4.2. Given that most cells within the matrix have at least one frequency hit across all data streams, we needed to identify the most salient challenges. To do this, we added the numbers that we used to gauge *strong* and *moderate* implications for each data stream together—i.e., 5, 3, and 2. Thus, in Figure 4.2., any challenge with ten or more frequency hits across all data streams was deemed to have *strong implications*, and any challenge with fewer than ten frequency hits was deemed to have *moderate implications*.

By using these groupings, we could better see which management levers needed to be pulled to address the thematic challenges. For example, this approach illuminated that organization and mission were the top two management areas implicated across all thematic issue areas, with personnel a close third. Moreover, this approach also illuminated multiple equipment concerns. While there were not many strongly implicated insights within the materiel management area, three of our thematic challenges had large frequency counts in this column. Importantly, equipment-related concerns were the second most-cited challenge across all data streams (29 mentions of old and antiquated equipment), only surpassed by posture concerns within the organization management area (36 mentions).

Appendix B. Additional Discussion of the Mission Demand-to-Sourcing Strategic Continuum

This appendix provides a more-detailed discussion of the Mission Demand-to-Sourcing Strategic Continuum shown in Figure 5.1. We begin with a description of each of the nodes.

- **Generate operational demands**—Most interviewees regardless of level or organizational positioning had to work to produce simple archives of what they had done in terms of missions. Demands for the AO community are joint, interagency, intergovernmental, and multinational, with different demands going to differing aspects of the community across active and component lines and by type of organization. Tracking of these demands by planners and mission overseers is a major first step in the community being able to better see itself. This is especially important because, as discussed using the matrix in Chapter 2, some in-competition and contingency demands on the AO career field are ill-defined or poorly communicated.
- **Receive and capture operational demands**—Once the demands are generated by the various entities in need, such as the Air Force, the joint warfighter community, or civil authorities for support, these demands need to be captured, triaged, and stored in a systematic fashion. These records will prove invaluable as the community is determining its current and future posturing and for those outside the community to better understand how to request and resource AO equities.
- **Identify available capability**—As with any good estimating process in mission analysis or the military decisionmaking process, understanding what capabilities exist to meet the demand requirements is a key step to successfully accomplishing missions, or at least mitigating risks by identifying shortfalls to be mitigated.
- **Determine capability gaps**—Once the triage of the demand process and identification of the resources available to meet mission demands is completed, the resulting product is sometimes a gap list to be mitigated. Mitigation strategies might include the use of contractors, reallocation of assets, or simply going without when feasible. Gap lists at the macro level of concern are key in determining which of the overall missions should be prioritized higher than others in an environment of constrained resources. This might result in higher-priority missions precluding the ability to provide resources to lower-priority missions—which could be necessary in managing a community that lacks the ability to meet all of its demands, as demonstrated by the hub-and-spoke examples in Chapter 5. Beyond meeting near-term operational demands, having a record of these gaps is crucial to demonstrating to the resources allocation community what needs to happen.
- **Source capability gaps**—The flow between what missions organizations are fulfilling is dynamic and somewhat well-known within the different stovepipes in the AO community; however, there is little understanding or awareness at even the highest levels of what the stovepipes are doing. Missions are being accomplished; however, information on how mission needs are met is not universally known or understood. Codifying how mission needs are met is an important input to posturing decisions for the community writ

large and for better mission awareness in the near term. Thus, the information needs to be captured and archived.

- **Organize, train, and equip**—Once the demands are understood, capabilities need to be organized from the various AO entities to meet the demands. This supply-based phase is very complex owing to the diversity in the types of units conducting specific or exclusive missions. Information awareness can be problematic because of the stovepiped nature of the community.
- **Execute the mission**—The culmination of phases at the macro and unit levels is the execution of the mission. Mission awareness and data tracking are in good order within most AO communities; however, we were not able to ascertain whether there is a proponent-level actor that has universal or macro-level awareness.

The Mission Demand-to-Sourcing Continuum model was developed in response to gaps in critical information encountered during interviews and based on activities that the proponent-level actors are currently trying to undertake to track and manage requirements and their corresponding fulfillment or reclama status. Several agencies in the field, as well as at the most strategic levels, could not tell us basic information like usage data and number of requests made, fulfilled, or those with reclamas. The inability to answer these questions was identified as a knowledge gap in the analysis but also presented the research team with the ability to build tools to conduct our analysis while simultaneously building products that could be used by AO practitioners to manage their equities in the future.

As an analytic baseline, the research team used the process map to assess the AO community and to provide the community with key insights as they set their vision and strategy bridging. The three key takeaways from our analysis, depicted in Figure B.1 and Table B.1, are as follows:

- The inability to clearly codify or capture the expanse of the demand history is a root cause of the AO community's inability to posture or align itself for current and future operational demands.
- The mission demands that can be seen right now are being completed, but at risk to efficiency and the future health of the community.
- Some of the overall mission is not accounted for, with the MAJCOMs currently requesting what they know is available as opposed to acknowledging operational needs—and subsequently filling some of those needs through other, at times less optimal, sourcing solutions.

Figure B.1 provides more-detailed results of an initial assessment that was shared with the vision and strategy bridging coordinators. Notably, none of the phases in the process map received a score of effective or efficient, with most phases needing attention and only two—identify available capability and mission execution—assessed as functional in their current states. Table B.1 details the specific rationale for determining the status of each phase in the continuum.

Figure B.1. Assessment Snapshot Using the Airfield Operations Mission Demand-to-Sourcing Continuum

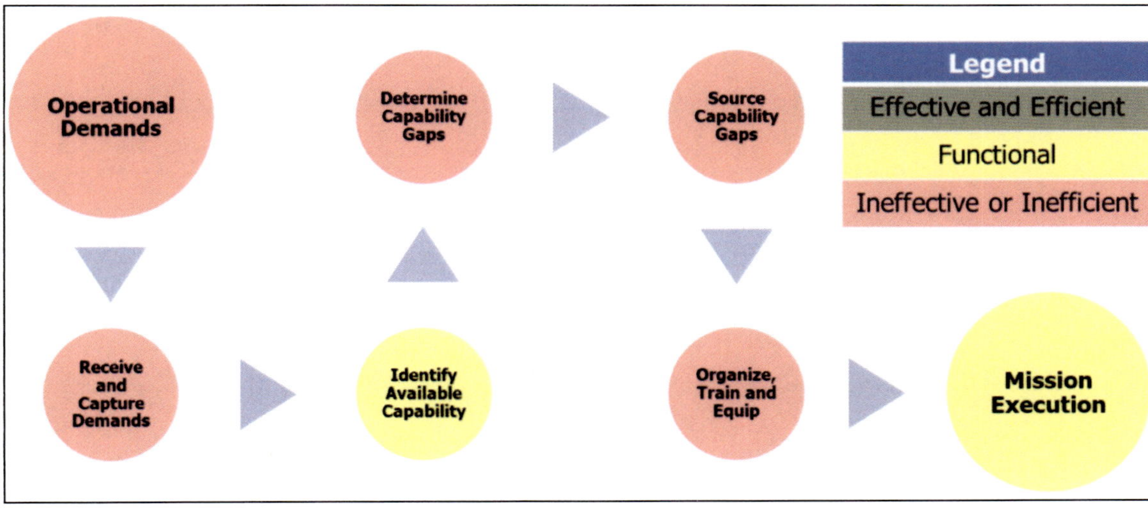

Table B.1. Description of the Assessment Snapshot Using the Airfield Operations Mission Demand-to-Sourcing Continuum

Airfield Operations Mission Demand-to-Sourcing Strategic Continuum			
Phases		Description/Needs	Status
I	Operational Demands	The missions, tasks, and operations from JIIM Partners and Homeland Security and set conditions for the size and composition of the forces to meet operational needs with minimal risk in effectiveness and efficiency.	Demands not captured based on availability shortfalls in the field in terms of gaps in authorized and/or on-hand people and equipment.
II	Receive and Capture Demands	• A multi-component systems is needed to process demands through an enterprise perspective. • NOTE: Capturing the demand history is necessary for validating adjustments to the force composition and design.	Inadequate demand history to validate requirements in the "job book" nor for force structure modifications.
III	Identify Available Capability	• Enterprise-wide availability and readiness picture needed across components to align capability with operational needs. • NOTE: Mission is getting done in less efficient manner based on shortages and lack of ability to validate increases.	Can see some of the availability and readiness picture and the picture is problematic based primarily on equipment and personnel shortfalls.
IV	Determine Capability Gaps	Gaps should be determined for the individual missions as well as at the enterprise capability level.	Gaps are known and worked around as opposed to resolved.
V	Source Capability Gaps	Gaps could be sourced a the enterprise or MAJCOM level with optimal efficiency coming from more strategic sourcing rather than less.	Little to strategic sourcing to account for known gaps and shortages.
VI	Organize, Train, and Equip	Organize, train, and equip the mission capability based on the operational demands based on validated history.	Incomplete history leading to lack of "job book" and appropriate force structure.
VII	Mission Execution	Execute the missions, tasks, and operations assigned as risk-free as possible in terms of mission accomplishment (effectiveness) and fiscal and toll on personnel (efficiency).	Getting the mission done with what is available; however, not efficiently, nor with the ability to justify increases to force structure

NOTE: JIIM = joint, interagency, intergovernmental, and multinational.

Appendix C. Data and Tools

This appendix describes the databases that were used for charts in this document that show AO career field composition, the AO tool, and the location maps. It also contains the R code for the AO model.[71]

Data Sources

Data were obtained from AFPC at Randolph AFB. These data sets are created at AFPC by extracting data from the Military Personnel Data System for several relevant groups. AFPC then scrubs these files for RAND and sends the latest versions each month (every quarter for the manpower and personnel accounting symbol tree files). The AFPC files used for this project were the following:

- Civilians–CAE
- Active Enlisted–AAE
- Active Officers–BAE
- Guard Enlisted–AGE
- Guard Officers–BGE
- Reserve Enlisted–ARE
- Reserve Officers–BRE
- Manpower Authorizations–MPW (includes all groups except Civilians)

Personnel from these files were selected if either their *primary* AFSC (PAFSC) or their *duty* AFSC (DAFSC) was one of the four AO AFSCs (13M, 1C1XX, 1C7XX, 1C8XX). If the PAFSC was missing, the duty DAFSC was used to include the individual in the database. In a few cases where PAFSC and DAFSC were missing, the secondary AFSC was used to identify an individual as part of the AO career field. In our analyses, we focused on personnel with a primary AFSC in AO.

The following status categories make up the "real inventory" of the personnel in our database, meaning that they are accountable as part of active-duty end strength after all non-relevant personnel (such as non–Air Force funded, deceased, prisoner of war) have already been filtered out.

- Permanent party—all personnel not in the other status categories; fully available/present for duty
- Students—current duty is being a student

[71] Data for this project was extracted by RAND colleague Perry Firoz, who provided this description of how the dataset was developed.

- Operational students—these students are rated personnel going to their initial full-time unit or back for requalification training. Test Pilot School is included in this category.
- Transients—personnel that are in transit (relocating to a new station or location, permanent change of station, temporary change of station, temporary duty, and others)
- Holdees—prisoners.

Analysis in this project focused on permanent party personnel.

The RAND AO Model

The RAND AO model uses as its data source an Excel spreadsheet with 26 columns. The first column has the name of the Air Force base and its location. Each of the other columns is an AO AFSC as follows:

- first column: base_f
- officer columns: 13M1, 13M3, 13M4
- enlisted ATC columns: 1C100, 1C111, 1C131, 1C151, 1C171, 1C191
- enlisted Airfield Management columns: 1C700, 1C711, 1C731, 1C751, 1C771, 1C791
- enlisted RAWS columns: 1C800, 1C813, 1C833, 1C853, 1C873, 1C891
- prefixed AFSCs for leadership, training, and other officer positions: B13M3, C13M3, R13M4, T13M3

R Code for RAND AO Model

```
library(tidyverse)
library(readxl)
library(stringr)
library(magrittr)
library(MASS)

setwd(“/Users/mrico/Desktop/AO Capabilities”)
list.files()

### READING IN FILES ###

filled_assigned_auth <- read_xlsx(“Draft comparison of authorized and assigned.xlsx”, sheet
= 2,
            skip =4, col_names = TRUE)

filled_assigned_auth$base <- unlist(lapply(str_split(filled_assigned_auth$base_f, “ \\(|\\)”),
‘[‘, 1))

filled_assigned_auth$type_of_base<- unlist(lapply(str_split(filled_assigned_auth$base_f, “
\\(|\\)”), ‘[‘, 2))
```

DEMAND FUNCTION

```
### This currently doesn't include contingency operations ###

calc_demand <- function(){
  majcom <- readline(prompt="Enter MAJCOM: ")
  hubs_num <- as.integer(readline(prompt="Enter number of hubs: "))
  spokes_num <- as.integer(readline(prompt="Enter number of spokes: "))
  airbridge_num <- as.integer(readline(prompt="Enter number of airbridge: "))

  if(majcom == "PACAF"|majcom == "USAFE"){
    ###################################
    ### PACAF CONEMP REQUIREMENTS ###
    ###################################
    airbridge <- c(0,4,4,2)
    hubs <- c(1,5,2,2)
    spokes <- c(0,3,1,1)

    standard <- data.frame(t(data.frame(hubs, spokes, airbridge)))
    names(standard) <-  c('13M','1C1XX','1C7XX','1C8XX')

    ### CALCULATING DEMAND
    demand <- data.frame(matrix(nrow = 4, ncol=4))
    names(demand) <- c('13M','1C1XX','1C7XX','1C8XX')
    demand[1,] = hubs_num*standard[1,]
    demand[2,] = spokes_num*standard[2,]
    demand[3,] = airbridge_num*standard[3,]
    demand[4,] = apply(demand, 2, function(x) sum(x, na.rm = TRUE))

    row.names(demand) <- c("Hubs", "Spokes", "Airbridge", "Total Demand")
    # print(paste("MAJCOM:", majcom))
    # print(paste("Number of hubs:", hubs_num))
    # print(paste("Number of spokes:", spokes_num))
    # print(paste("Number of airbridge:", airbridge_num))

  return(demand)
  }
}

calc_demand()
```

SUPPLY FUNCTION

```
calc_supply <- function(majcom, include_korea, deployable, fraction_of_deployable,
majcom_fraction){
```

```
majcom <- readline(prompt = ("MAJCOM:"))

avail_to_deploy = filled_assigned_auth %>%
  filter(type_of_base == majcom)

avail_to_deploy$Total_13M <-
rowSums(avail_to_deploy[,str_detect(names(avail_to_deploy), "(13M)")], na.rm = T)
avail_to_deploy$Total_1C1 <-
rowSums(avail_to_deploy[,str_detect(names(avail_to_deploy), "^(1C1)")], na.rm = T)
avail_to_deploy$Total_1C7 <-
rowSums(avail_to_deploy[,str_detect(names(avail_to_deploy), "^(1C7)")], na.rm = T)
avail_to_deploy$Total_1C8 <-
rowSums(avail_to_deploy[,str_detect(names(avail_to_deploy), "^(1C8)")], na.rm = T)

avail_to_deploy$korea <- c("no","no","no","no","no", "yes", "no", "yes", "no")
avail_to_deploy$deployable <- c("no","yes","yes","no","no", "no", "no", "no", "no")

all_PACAF_supply = avail_to_deploy %>%
   dplyr::select(base, Total_13M,Total_1C1,Total_1C7, Total_1C8, korea, deployable)

include_korea <- readline(prompt = "Should we include Korea-based bases? (YES/NO)")
 if(include_korea == "NO"){
   x <- all_PACAF_supply %>%
    filter(korea != 'yes')
   }
 else{
   x <- all_PACAF_supply
 }
### IF INCLUDING KOREA BASES, THEY FALL INTO NON-DEPLOYABLE BASE
CATEGORY

nondeployable <- readline(prompt = "Do you only want to include non-deployable bases?
(YES/NO)")
 if(nondeployable == "NO"){
  x %<>%
    filter(deployable == "yes")
 fraction_of_nondeployable = as.numeric(0)
 }
 else{
   fraction_of_nondeployable <- as.numeric(readline(prompt = "At what percentage would
you like to include the non-deployables? (This will be across the entire MAJCOM)"))
 }
fraction_of_deployable <- as.numeric(readline(prompt = "At what percentage would you
like to include deployable personnel?"))

x %<>%
```

```
    dplyr::mutate(fraction = case_when(
      x$deployable == "yes" ~ fraction_of_deployable,
      x$deployable == "no" ~ fraction_of_nondeployable
      ))
  x %<>%
    mutate(across(2:5, ~ .x *fraction))

  x[,2:5] <- floor(x[,2:5])

return(x)
}

calc_supply()

calc_demand_shortfall <- function(){
  d <<- calc_demand()
  s <<- calc_supply()
  ds <<-  colSums(s[,2:5]) – d[nrow(d),]
  rownames(ds) <<- 'Total Supply Shortfall'

  ds <<- as.data.frame(t(ds))

  plot <- ds %>%
    mutate(Color = ifelse(`Total Supply Shortfall` < 0, "red","green")) %>%
    ggplot(aes(x = rownames(ds) ,y = `Total Supply Shortfall`, fill = Color))+
    geom_col()+
    scale_fill_identity(guide = FALSE) +
    labs(x = '')

  return(list(ds, plot))
}

calc_demand_shortfall()
```

Abbreviations

AB	Air Base
ACC	Air Combat Command
ACE	Agile Combat Employment
AEF	air and space (or aerospace) expeditionary force
AFB	Air Force Base
AFFORGEN	Air Force Force Generation (model)
AFFSA	Air Force Flight Standards Agency
AFI	Air Force Instruction
AFMAN	Air Force Manual
AFNORTH	Air Force Northern Command
AFPC	U.S. Air Force Personnel Center
AFPD	Air Force Policy Directive
AFR	Air Force Reserve
AFSC	Air Force specialty code
AFTTP	Air Force Tactics, Techniques, and Procedures
AMC	Air Mobility Command
ANG	Air National Guard
AO	airfield operations
AOR	area of responsibility
ATC	air traffic control
ATCALS	air traffic control and landing systems
ATCS	Air Traffic Control Squadron
BTF	Bomber Task Force
COCOM	combatant command
CONEMP	concept of employment
CONOP	concept of operation
CONUS	continental United States
CRG	Contingency Response Group
CRS	Contingency Response Squadron
DFE	Dynamic Force Employment
DoD	Department of Defense
DOTmLPF-P	doctrine, organization, training, materiel, leadership and education, personnel, facilities, and policy
DRRS-S	Defense Readiness Reporting System–Strategic
FAA	Federal Aviation Administration

FCT	FAA Contract Tower
FOB	forward operating base
FOS	forward operating site
HADR	humanitarian and disaster relief
JBMDL	Joint Base McGuire-Dix-Lakehurst
MAJCOM	major command
MCA	multi-capable airman/airmen
MOB	main operating base
NDS	National Defense Strategy
NSS	National Security Strategy
OCONUS	outside CONUS
OJT	on-the-job training
OSS	operations support squadron
PACAF	Pacific Air Forces
PAF	(RAND) Project Air Force
RAWS	radar, airfield, and weather systems
RMC	Regional Maintenance Center
SME	subject-matter expert
TACAN	tactical air navigation
USAF	U.S. Air Force
USAFE	U.S. Air Forces in Europe
USCENTCOM	U.S. Central Command
UTC	unit type code
VOR	Very High Frequency Omnidirectional Range
VORTAC	VOR/Tactical Air Navigation

References

423rd Mobility Training Squadron, *Contingency Airfield Operations*, Air Force Tactics, Techniques, and Procedures (AFTTP) 3-4.4, September 30, 2017.

53rd Air Traffic Control Squadron, "53rd Air Traffic Control Squadron Continues Mission in Spite of Pandemic," Robins Air Force Base, June 15, 2020. As of April 13, 2022: https://www.robins.af.mil/News/Article-Display/Article/2220051/53rd-air-traffic-control-squadron-continues-mission-in-spite-of-pandemic/

AFI—*See* Air Force Instruction.

AFMAN—*See* Air Force Manual.

AFPD—*See* Air Force Policy Directive.

Air Force Doctrine Publication 4-0, *Combat Support*, U.S. Air Force, January 5, 2020.

Air Force Instruction 36-2101, *Classifying Military Personnel (Officer and Enlisted)*, Department of the Air Force, March 7, 2006.

Air Force Instruction 38-101, *Air Force Organization*, Department of the Air Force, March 16, 2011.

Air Force Instruction 10-601, *Operational Capability Requirements Development*, Department of the Air Force, November 6, 2013.

Air Force Instruction 11-201, *Flying Operations*, Department of the Air Force, November 30, 2018.

Air Force Instruction 10-202, *Contingency Response Forces*, Department of the Air Force, June 11, 2020.

Air Force Instruction 36-2670_ AFGM2020-03, *Air Force Guidance Memorandum (AFGM) to AFI 36-2670, Total Force Development*, Department of the Air Force, December 7, 2020.

Air Force Instruction 10-201, *Force Readiness Reporting*, Department of the Air Force, December 22, 2020.

Air Force Manual 38-208, Vol. 1, Air Force Management Engineering Program (MEP) – Processes, U.S. Air Force, March 29, 2002.

Air Force Manual 38-102, Manpower and Organization Standard Work Processes and Procedures, U.S. Air Force, September 4, 2019.

Air Force Manual 13-204, Vol. 1, *Management of Airfield Operations*, U.S. Air Force, July 22, 2020a.

Air Force Manual 13-204, Vol. 4, *Radar, Airfield and Weather Systems*, U.S. Air Force, July 22, 2020b.

Air Force Policy Directive 13-2, *Air Traffic Control, Airfield, Airspace, and Range Management*, U.S. Air Force, January 3, 2019.

Air Force Tactics, Techniques, and Procedures, Vol. 3-4.4, *Contingency Airfield Operations*, November 2, 2017.

Air National Guard, "Air Traffic Control Mission," webpage, undated. As of September 28, 2021: https://www.goang.com/discover-ang/missions/ground-support/air-traffic-control-mission.html

ANG—*See* Air National Guard.

ATCS—*See* Air Traffic Control Squadron.

Benson, Lawrence R., *USAF Aircraft Basing in Europe, North Africa, and the Middle East: 1945–1980*, Office of History, Headquarters, United States Air Forces in Europe, Ramstein Air Base, Germany, April 23, 1981, Declassified May 28, 2010.

Biden, President Joseph R., Jr., *Interim National Security Strategic Guidance*, Washington, D.C.: The White House, March 2021.

Brown, Senior Airman Jason J., "AER Next Represents Next Evolution in Air Force Expeditionary Operations," Joint Base Langley-Eustis News, May 12, 2014. As of April 13, 2022: https://www.jble.af.mil/News/Article-Display/Article/844091/aef-next-represents-next-evolution-in-air-force-expeditionary-operations

Cohen, Rachel, "New 'Lead Wing' Deployment Plan for Combat Aircraft Is Being Tested, Refined," *Air Force Times*, May 25, 2021. As of April 13, 2022: https://www.airforcetimes.com/news/your-air-force/2021/05/25/new-lead-wing-deployment-plan-for-combat-aircraft-is-tested-refined/

Conley, Raymond E., Albert A. Robbert, Joseph G. Bolten, Manuel Carrillo, and Hugh G. Massey, *Maintaining the Balance Between Manpower, Skill Levels, and PERSTEMPO*, Santa Monica, Calif.: RAND Corporation, MG-492-AF, 2006. As of April 13, 2022: https://www.rand.org/pubs/monographs/MG492.html

Corbett, Julian, *Some Principles of Maritime Strategy*, Scotts Valley, Calif.: The Perfect Library, 1911.

Dahlman, Carl J., Robert Kerchner, and David E. Thaler, *Setting Requirements for Maintenance Manpower in the U.S. Air Force*, Santa Monica, Calif.: RAND Corporation, MR-1436-AF, 2002. As of April 13, 2022: https://www.rand.org/pubs/monograph_reports/MR1436.html

DeBlois, Bradley, Thomas Light, Daniel M. Romano, Michael Boito, John G. Drew, Paul Emslie, Michael Kennedy, Kathryn O'Connor, and Jonathan William Welburn, *Options for Enhancing the Effectiveness of Maintenance Force Structure: Examining the Decline in KC-135 Availability*, Santa Monica, Calif.: RAND Corporation, 2020, Not available to the general public.

Department of the Air Force, Pacific Air Forces, "Agile Combat Employment (ACE): PACAF Annex to Department of the Air Force Adaptive Operations in Contested Environments," June 2020.

Esper, Mark T., Secretary of Defense, *Implementing the National Defense Strategy: A Year of Successes*, July 17, 2020. As of September 14, 2021: https://media.defense.gov/2020/Jul/17/2002459291/-1/-1/1/NDS-FIRST-YEAR-ACCOMPLISHMENTS-FINAL.pdf

FAA—*See* Federal Aviation Administration.

Federal Aviation Administration, *Emergency Security Control of Air Traffic (ESCAT)*, Advisory Circular (AC) No. 99-1E, August 8, 2018.

Federal Aviation Administration, *Air Traffic Organization Policy,* FAA Order JO 7110.65Y, August 15, 2019.

Federal Aviation Administration, "FAA Contract Tower Programs," webpage, February 26, 2020. As of September 28, 2021: https://www.faa.gov/about/office_org/headquarters_offices/ato/service_units/mission_support/faa_contract_tower_program/

Headquarters Air Force Personnel Center (AFPC/DP3DW), *Air Force Enlisted Classification Directory (AFECD): The Official Guide to the Air Force Enlisted Classification Codes*, April 30, 2020.

Joint Staff, *Description of the National Military Strategy 2018*, Washington, D.C., 2018. As of April 13, 2022: https://www.jcs.mil/Portals/36/Documents/Publications/UNCLASS_2018_National_Military_Strategy_Description.pdf

Levine, Peter, Performing the Duties of the Under Secretary of Defense for Personnel and Readiness, *Directive-type Memorandum (DTM)-17-004—Department of Defense Expeditionary Civilian Workforce*, January 25, 2017, Incorporating Change 4, Effective April 19, 2021.

Lynch, Kristin F., Katherine C. Hastings, Danielle C. Tarraf, Jonathan L. Brosmer, John G. Drew, Eli M. Kohlenberg, Ronald G. McGarvey, Amaka Oji, Eric Perinovic, and Matthew Sargent, *Persistent Logistics: An Evaluation of Strategies and Posture Options for Maintenance: A Case Study in the Pacific Theater*, Santa Monica, Calif.: RAND Corporation, 2022, Not available to the general public.

Mattis, James, *Summary of the 2018 National Defense Strategy of the United States of America: Sharpening the American Military's Competitive Edge*, Washington, D.C.: U.S. Department of Defense, 2018.

Mills, Patrick, John G. Drew, John A. Ausink, Daniel M. Romano, and Rachel Costello, *Balancing Agile Combat Support Manpower to Better Meet the Future Security Environment*, Santa Monica, Calif.: RAND Corporation, RR-337-AF, 2014. As of April 13, 2022: https://www.rand.org/pubs/research_reports/RR337.html

Mills, Patrick, James A. Leftwich, John G. Drew, Daniel P. Felten, Josh Girardini, John P. Godges, Michael J. Lostumbo, Anu Narayanan, Kristin Van Abel, Jonathan Welburn, and Anna Jean Wirth, *Building Agile Combat Support Competencies to Enable Evolving Adaptive Basing Concepts*, Santa Monica, Calif.: RAND Corporation, RR-4200-AF, 2020. As of September 28, 2021: https://www.rand.org/pubs/research_reports/RR4200.html

Mineau, David A., Director, Current Operations (AF/A2O), *2020 Airfield Operations (AO) Executive Council Minutes*, Washington, D.C., January 6, 2021.

National Guard Bureau, *ANG ATCS 101*, PowerPoint briefing prepared for the RAND Corporation, May 2021.

Nolte, Anna, "Multi-Capable Airmen Course Innovates Airpower Capabilities," Pacific Air Forces, February 26, 2021. As of September 28, 2021: https://www.pacaf.af.mil/News/Article-Display/Article/2518090/multi-capable-airmen-course-innovates-airpower-capabilities

O'Donohue, Daniel J., Director, Joint Force Development, Joint Chiefs of Staff (JCS), *Competition Continuum*, Joint Doctrine Note 1-19, Washington, D.C., June 3, 2019.

PACAF—*See* Pacific Air Forces.

Pacific Air Forces, *Airfield Operations: Agile Combat Employment (ACE) CONCEPT OF OPERATIONS (CONOPS)*, Joint Base Pearl Harbor–Hickam, Hawaii, December 2019.

Reiter, Thomas E., *USAF Aircraft Maintenance Organizational Structure: Where We've Been, Where We Are, What's the Future*, research report, Maxwell Air Force Base, Ala.: Air War College, Air University, April 1988.

Robbert, Albert A., Lisa M. Harrington, Louis T. Mariano, Susan A. Resetar, David Schulker, John S. Crown, Paul Emslie, Sean Mann, and Gary Massey, *Air Force Manpower Determinants: Options for More-Responsive Processes*, Santa Monica, Calif.: RAND Corporation, RR 4420-AF, 2020. As of April 13, 2022: https://www.rand.org/pubs/research_reports/RR4420.html

Robbert, Albert A., Lisa M. Harrington, Tara L. Terry, and Hugh G. Massey, *Air Force Manpower Requirements and Component Mix: A Focus on Agile Combat Support*, Santa Monica, Calif.: RAND Corporation, RR-617-AF, 2014. As of April 13, 2022: https://www.rand.org/pubs/research_reports/RR617.html

Roberts, James B., Director, Programs and Requirements, Air Force Flight Standards Agency, "NAVAID Status CAO," briefing, February 8, 2021, Not available to the general public.

Roth, John P., Acting Secretary of the Air Force; Charles Q. Brown, Jr.,Chief of Staff, U.S. Air Force; and John W. Raymond, Chief of Space Operations, U.S. Space Force, *Department of the Air Force Posture Statement Fiscal Year 2022*, Department of the Air Force presentation to the Committees and Subcommittees of the U.S. Senate and the House of Representatives, 1st Session, 117th Congress, prepared testimony released May 7, 2021.

Salazar, Heather, "B-2 Spirit Integrates with NATO Ally at Lajes Field," *Air Force News*, March 25, 2021. As of July 13, 2021: https://www.af.mil/News/Article-Display/Article/2547730/b-2-spirit-integrates-with-nato-ally-at-lajes-field/

Schmidt, Eric, "36 Contingency Response Group: Afld Ops UTC and MCA," briefing presented at October 2020 Executive Council, 2020.

Snyder, Don, Patrick Mills, Manuel Carrillo, and Adam Resnick, *Supporting Air and Space Expeditionary Forces: Capabilities and Sustainability of Air and Space Expeditionary Forces*, Santa Monica, Calif.: RAND Corporation, MG-303-AF, 2006. As of April 13, 2022: https://www.rand.org/pubs/monographs/MG303.html

Trump, Donald J., President of the United States, *National Security Strategy of the United States of America*, Washington, D.C., December 2017.

U.S. Air Force, "Capability-Based Manpower Standard 13ex00," Airfield Operations Flight, September 24, 2007.

U.S. Air Force, "Air Force Personnel Center," fact sheet, March 27, 2015. As of July 26, 2021: https://www.af.mil/About-Us/Fact-Sheets/Display/Article/104554/air-force-personnel-center/

U.S. Air Force, *Career Field Education and Training Plan (CFETP), AFSC 13MX, Airfield Operations Officer, Parts 1 and 2*, Washington, D.C.: Department of the Air Force, Headquarters U.S. Air Force, June 21, 2016.

U.S. Air Force, *Career Field Education and Training Plan (CFETP), AFSC 1C8X3, Radar, Airfield & Weather Systems (RAWS)*, Parts I and II, Washington, D.C.: Department of the Air Force, Headquarters U.S. Air Force, February 6, 2018 (Corrective Actions applied on May 22, 2018).

U.S. Air Force, "Department of the Air Force Guidance Memorandum to DAFI 10-401, Air Force Operations Planning and Execution," DAFI10-401_DAFGM2021-01, January 19, 2021a.

U.S. Air Force, "New Force Generation Model Builds High-end Readiness, Sustainability for Joint Force," press release, August 5, 2021b. As of May 3, 2022: https://www.af.mil/News/Article-Display/Article/2720333/new-force-generation-model-builds-high-end-readiness-sustainability-for-joint-f/

U.S. Joint Chiefs of Staff, *Joint Operations*, Joint Publication 3-0, Washington, D.C., January 17, 2017, incorporating change 1, October 22, 2018.

U.S. Joint Chiefs of Staff, *Contingency Basing*, Joint Publication 4-04, Washington, D.C., January 4, 2019.

Wilson, Billy R., "Airfield Operations Personnel Discussion," briefing at Airfield Operations Executive Council meeting, October 2020.

Wilson, Billy R., *Air Force Airfield Operations Vision and Bridging Strategy 2021–2035*, DRAFT version 2.8, June 2021, Not available to the general public.

Wilson, Bryan, Radar, Airfield and Weather Systems Career Field Manager, "Future RAWS Warfighting Concepts: Gap Validation & OPR Assignment," briefing at Airfield Operations Executive Council meeting, October 2020.

Wirth, Anna Jean, Thomas Light, Daniel M. Romano, Shane Tierney, Ronald G. McGarvey, Moon Kim, Michael J. Lostumbo, Amanda Nguyen, Paul Emslie, and John G. Drew, *Evaluating* Alternative *Maintenance Manpower Force Structure Concepts for the F-35A*, Santa Monica, Calif.: RAND Corporation, RR-4433-AF, 2020. As of April 13, 2022: https://www.rand.org/pubs/research_reports/RR4433.html

Wright, Jessica L., Acting Under Secretary of Defense (Personnel and Readiness), "Under Secretary of Defense (Personnel & Readiness) Deployment-to-Dwell, Mobilization-to-Dwell Policy Revision," memorandum for Secretaries of the Military Departments, Chairman of the Joint Chiefs of Staff, November 1, 2013.